Biodiversity of Ecosystems

Edited by Levente Hufnagel

Published in London, United Kingdom

IntechOpen

Supporting open minds since 2005

Biodiversity of Ecosystems
http://dx.doi.org/10.5772/intechopen.94786
Edited by Levente Hufnagel

Contributors
Hana Tamrat Gebirehiwot, Alemayehu Abera Kidanu, Megersa Tafesse Adugna, Menoh A. Ngon René, Tsoata
Esaïe, Tsouga Manga Milie Lionelle, Owona Ndongo Pierre-André, Vo Thi Ngoc My, Nguyen Van Thanh,
María Del Pilar Rodríguez-Guzmán, Mulugeta Aytenew, Ishwari Singh Bisht, Jai Chand Rana, Sarah Jones,
Rashmi Yadav, Natalia Estrada-Carmona, Richard E. Rice, Hassanali Mollashahi, Magdalena Szymura,
Prerna Gupta, Sadhna Tamot, Wael Mahmoud ElSayed, Shahenda Abu ElEla Ali Abu ElEla, Nakamura Koji,
Noelia Tourón, Estefanía Paredes, Damián Costas, Levente Hufnagel, Ferenc Mics

Notice
Statements and opinions expressed in the chapters are these of the individual contributors and not
necessarily those of the editors or publisher. No responsibility is accepted for the accuracy of
information contained in the published chapters. The publisher assumes no responsibility for any
damage or injury to persons or property arising out of the use of any materials, instructions, methods
or ideas contained in the book.

First published in London, United Kingdom, 2022 by IntechOpen
IntechOpen is the global imprint of INTECHOPEN LIMITED, registered in England and Wales,
registration number: 11086078, 5 Princes Gate Court, London, SW7 2QJ, United Kingdom
Printed in Croatia

British Library Cataloguing-in-Publication Data
A catalogue record for this book is available from the British Library

Additional hard and PDF copies can be obtained from orders@intechopen.com

Biodiversity of Ecosystems
Edited by Levente Hufnagel
p. cm.
Print ISBN 978-1-83969-487-5
Online ISBN 978-1-83969-488-2
eBook (PDF) ISBN 978-1-83969-489-9

We are IntechOpen,
the world's leading publisher of
Open Access books
Built by scientists, for scientists

6,000+
Open access books available

146,000+
International authors and editors

185M+
Downloads

Our authors are among the

156
Countries delivered to

Top 1%
most cited scientists

12.2%
Contributors from top 500 universities

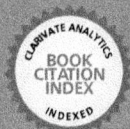

Interested in publishing with us?
Contact book.department@intechopen.com

Meet the editor

Dr. Levente Hufnagel is an associate professor and the head of the Research Institute of Multidisciplinary Ecotheology, John Wesley Theological College, Hungary, working on ecology, biogeography, ecological research methodology, ecotheology, and sustainability. He obtained a master's degree in Ecology and Evolutionary Biology and a Ph.D. in Hydrobiology from Eötvös Loránd University, Hungary. He also has a Ph.D. in Agricultural Science from Szent István University, Hungary, as well as other degrees from the Corvinus University of Budapest and Adventist Theological College. He has more than twenty years of experience leading Hungarian academic institutions and teaching and supervising Ph.D., MSc, and BSc students from various social and cultural backgrounds. He has more than 240 scientific publications (on both aquatic and terrestrial ecological aspects of plants, animals, and microbes at the community as well as population levels) and more than 1000 independent citations to his credit. As a participant in several big ecological research and development projects, Dr. Hufnagel has significant experience in multidisciplinary collaborations with more than 200 coauthors in different publications. He is also editor-in-chief of an international scientific journal.

Contents

Preface

Humanity, together with its global society, culture, and economy, is only a part of the earth's biosphere as a complex ecological system. The existence of humanity depends on the functioning of the biosphere (e.g., climate regulation, self-purification, ecosystem services) and the health of ecosystems. The state of ecosystems is determined not only by the biomass contained in them and their biological activity and productivity but also by the functional redundancy that results in reliability and a high degree of biodiversity. Maintaining the biodiversity of ecosystems (and ultimately our planet) is in the common interest of people worldwide. However, the conservation of biodiversity requires not only national parks, protected species, and conservation biological efforts, but also our economy, our political life, the organization of our cities, and the way we operate agriculture. The success of these efforts depends on the basic and applied research outlined in this book. It discusses a variety of topics, including abiotic factors that affect biodiversity, conservation and sustainability efforts, and urban and agricultural ecosystems. Chapters include case studies to illustrate special methodical problems and research approaches in the field.

This book is useful for researchers, lecturers, students, and other interested readers who would like to get some insight into the biodiversity research of human-influenced ecosystems.

Levente Hufnagel
Research Institute of Multidisciplinary Ecotheology,
John Wesley Theological College,
Budapest, Hungary

Section 1

Pure Ecology and Conservation

Introductory Chapter: Factors That Affect Biodiversity and Species Richness of Ecosystems - A Review

Levente Hufnagel and Ferenc Mics

1. Introduction

The general latitudinal trend is that species diversity declines as latitude increases [1]. This appears to be the case for almost all terrestrial plants and animals [2, 3]. It is also usually used for the distribution of marine species [4, 5]. This latitudinal pattern in species richness is detectable in different spatial scales, habitats, and taxonomic groups [6]. However, because of the lack of agreement on the dynamics of site-to-site heterogeneity in species composition (b-diversity) through latitudinal gradients, latitudinal variations in species co-occurrence remain a central issue in ecology [7, 8]. Living species colonized and changed nearly all aquatic and terrestrial ecosystems on Earth, and thereby developed in their many shapes, physiology, and life history from origins more than three billion years ago. Ecosystems are not uniformly rich in species but this richness shows pattern across the world, and observers have long been puzzled about the origins of striking diversity trends such as the latitudinal diversity gradient, elevational gradients in terrestrial ecosystems, and bathymetric gradients in the sea. Since Darwin and Wallace, biologists have been focused on elucidating the mechanisms that lead to diversity. To explain why some regions of the world host greater numbers of taxa than other areas, studies on lineage differentiation and longevity (evolutionary biology) and organismal survivorship and coexistence (ecology, etc.) must be combined. In the last decade or so, new data on the pattern of biodiversity, fine-scale maps of climate and environmental factors, and huge developments in the reconstruction of the tree of life have been ongoing. Biogeographical and phylogenetic data integrations also restructured the patterns of the global distribution of species diversity as well as phylogenetic connections of organisms on various taxonomic scales. The processes that underlie species diversity trends are crucial to study, particularly when people influence important environmental changes [9]. Humans are diminishing habitat space for the majority of the world's organisms, while global temperature and precipitation regimes are shifting, and these factors are in the focus of many hypotheses describing the nature and maintenance of species diversity. To complicate things, some of these assumptions involve the same factors, namely geographic area, energy climate stability, and biotic interactions, and generate similar predictions about diversity and the rate of speciation because most of the hypothetical factors coincide with the latitude. However, the intense theoretical and methodological attention focused on species richness gradients has not been applied to consider the broadscale distribution of other essential components of biodiversity, such as intraspecific diversity. Intraspecific diversity—whether defined as functional diversity, phylogenetic diversity, population richness, or genetic diversity within and among populations—can influence species' geographic distributions and responses to environmental change [10–13], community structure, and ecosystem functioning [14, 15]. Given the alarming pace of species loss and human modification of natural ecosystems, the importance to quantify biodiversity is crucial (e.g., [16–20]). Species richness, namely the number of species per unit area, is perhaps the simplest and most often used indicator of biological diversity. A substantial amount of biological research has been done using species richness as a metric to explain what causes, and is caused by, biodiversity. Species richness is often used as a measure of diversity within a single biological community, ecosystem, or habitat (e.g., [21]). Pianka [22] wrote the first review paper on large-scale diversity gradients and reviewed major hypotheses to explain the latitudinal diversity gradient. Latitudinal gradients have been known for almost a century in species diversity, and nowadays, some of these polar-equatorial patterns have been explored in depth and several authors added new hypotheses to explain latitudinal gradients [8, 23, 24]. The gradient-forming drivers are differentiated on the basis of significance to different stages of lineage divarication, survival, and allopatric; parapatric and sympatric speciation can be integrated into this framework [25].

2. Abiotic factors

When a population of a certain taxon extends, its range to an extra-tropical area has to face evolutionary trade-offs that influence speciation. Cold tolerance is the most important trait to avoid frost damage poleward from frost line, in the cold season species without frost tolerance can be easily eliminated, even if freezing does not occur every year. Significant energy must be invested in frost resistance at the expanse of growth and reproduction [26]. Thus, they cannot grow and produce offsprings as fast as frost-intolerant species and are out-competed in frost-free areas [27]. The breadth of tropical climate zone and physiological pressures combined with biotic factors hinder northward wandering and this phenomenon is true for plants and animals (termed Dobzhansky-MacArthur phenomenon) [28]. Animals have to struggle with frost, but they are more mobile and can migrate or become dormant in winter, so this trade-off is not pronounced, but vertebrate and invertebrate fauna almost completely turns over northward and southward moving away from the equator [29]. In addition, many animals' distribution coincides with that of feeding plants [30].

Many taxonomic groups are of tropical origins, and the main source of diversity is the tropical region (such as birds, amphibians, angiosperms, and many marine invertebrates) [23, 31–33]. Terrestrial phylogenetical lineages' expansion is restricted by physiological tolerance toward cold or arid environment, evolution of adaptation is particularly difficult, and these traits have evolved fairly infrequently [34]. Hence, most speciation events occur within the border of biomes to lineages that do not cross boundaries of bioregion. Bacteria do not show particulate biome adherence [35], and criteria that link biodiversity to bioregions are not confirmed; hence, the prevalence of biome crossing lineages, dispersal rates of lineages, and the size of transition zones between biomes should be determined [36]. On the regional scale, speciation is believed to be affected by the population size with the presumption that bigger areas support a larger population [37]. Population size, in turn, is positively related to the physical extent of the bioregion and genetic diversity. Size of distribution area is believed positively correlated with allopatric speciation, and small-ranged species have no opportunity to be divided by a barrier within the range. If ranges are large enough to surround a barrier, the species has also less chance to be divided into two new species. Leading to conclusion that medium-ranged species are predicted to have the highest speciation rate [38]. Bioregions with tectonic activity such as mountain lifting have improved opportunities for speciation and environmental gradient by emerging new mountain ranges. Finally, wide population ranges result in great genetic and phenotypic variability increasing the possibility of survival in a highly heterogenic environment, which is a prerequisite for parapatric speciation [39]. There is a clear correlation between the extent of the area and the likelihood of extinction. In smaller bioregions, the average range of species is also smaller. This is associated with a higher likelihood of extinction due to catastrophic events [40]. In addition, in small populations, both genetic and phenotypic variations are small, which easily leads to inbreeding, and are therefore less able to adapt to changing environmental conditions [41].

The extent and position of the temperate belt have changed many times in the history of the Earth, to which glaciations have contributed [42]. Of course, temperate bioregions have existed for at least 50 million years, where mammals, birds, reptiles, amphibians, and seed plants also lived [43].

The size of the population also positively correlated with the amount of available energy. Of course, in a larger bioregion and in a larger population, this is even more pronounced [44]. This is analogous to the idea of population abundances from metabolic theory, and this assumes that more available energy in one region allows for a larger population size, affecting speciation and extinction rate [45]. High-productivity areas tend to be more heterogeneous, with several sources available to populations, than low-productivity bioregions, increasing the likelihood of speciation [46]. It should be mentioned that energy as a factor influencing speciation rate is not equal to the theory that the total energy available limits the number of species in a given region. Global trends of annual net primary production (NPP) of natural biomes are critical to understanding global (natural and anthropogenic) carbon budgets, and they are essential to understanding the adaptive relationships and the evolution of ecosystems and biomes. The competitive exclusion hypothesis is one of many hypotheses that seek to understand biodiversity dynamics by focusing on primary productivity [47], the energy-richness (or more individuals) hypothesis [48], the integrated evolutionary speed hypothesis [49], and the biological relativity to water-energy dynamics hypothesis [50, 51]. The first of these theories is based on the premise that the prevailing relationship between productivity and species diversity is unimodal (or hump-shaped), while the other three are based on the assumption that the predominant relationship is positive. Another different perspective for explaining the factors that ensure species coexistence is the hypothesis that there is a carrying capacity determined by the energy available. The ecoregions attain equilibrium over time in terms of the number of species and the speciation and extinction are balanced. The addition of a new species to the area causes a decrease in the population size of the previous species, increasing the chance of extinction because there is a limit on the number of individuals for each species. New arrival may lead to extinction of a local species [52]. The ecological needs of different species differ, which concludes that there is more competition between closely related species because they have similar ecological needs. Over time, diversification slows down resulting in a pattern of diversity-dependent diversification [53]. The assumption is that nearby species compete for available resources in a zero-sum game, and the ecological limit determines the maximum number of species within a clade in a region. During the Cenozoic period, when the climate became cooler, speciation also slowed down, as lower average temperatures also lowered net primary production. In the case of a lower net primary production, the number of new species that could theoretically be formed is also smaller [54]. In the case of higher net primary production, more species may develop but a slowdown can be observed here as well, reaching a limit.

A number of phylogenetic studies have reported diversity-dependent species formation, but in many cases due to methodological and sampling problems, in the practice, it has failed to demonstrate that the evolution of new species would slow down after reaching a theoretical limit [55].

Machac and Graham [56], on the other hand, found in their study that the formation of new species does not slow down in the tropical region and that the slowdown may be an artifact and there is no real limit to species formation.

This does not mean that co-occurring species' biological traits do not impact their diversification rates. Several examples have been reported that how the evolution of new species is influenced by species interactions [57]. However, it is very difficult to detect the persisting and speciation-promoting effect of the area and the source on clade-level properties in lineages and to distinguish it from the diversity-limiting effect of area and productivity [58].

In addition, the theory of ecological limitations did not provide a strong conceptual relation between how the volume of resources is associated with carrying capacity for all organisms, and how resources could then be specifically related to the diversity of species [59]. Large-scale dynamics of species diversity did not correlate well with numbers of organisms, according to previous reviews that found little evidence for the species-energy hypothesis [60]. While new immigration to an area (or new species emerging in situ) would inevitably consume some of the region's available resources, it is unclear how a new species would affect the community abundances of all other native species unless only one other species exists in the area. The relative abundances of any of these species will differ in an infinite number of ways given a certain amount of resources and a certain degree of species diversity [61]. Furthermore, observational experiments have shown that common organisms in an area may have extraordinarily high relative abundances [62]. If newcomer species decrease the abundance of common species while not reducing the abundance of rarer species, so more rare species can be easily accommodated within the field. Our knowledge of how the relative abundance of species changes over time is complicated by the Milankovitch cycle, which also changes the extent and shape of bioregions in 14, 21, and 100,000-year cycles. These periods are often shorter than we would expect for species formation or extinction [63, 64]. Furthermore, with their recent histories, certain bioregions such as tundra and boreal bioregions are unlikely to have reached equilibrium diversity for trees or vertebrates [65].

Finally, it is unclear which scientific evidence would be required to assess whether bioregions have attained their carrying potential for species diversity. Rather than thinking about whether or not a bioregion represents an equilibrium or nonequilibrium structure, I believe it is more efficient to concentrate on how various drivers can affect speciation and extinction [66].

Species energy, time-integrated area, and tropical conservatism hypotheses are based on some mechanistic assumptions, taking into account population size, speciation, and extinction rates, these are worth combining. These concepts are based on the fact that by observing area and energy over a long period of time, diversity and diversification rates can be estimated on a large scale. Marin et al. [67] concluded that one of the most important factors in predicting species richness is a time-integrated area (area through time). Ecological and climatic stability influences species richness indirectly, altering the evolutionary time (i.e., persistence time) and rate. According to their discovery, global heterogeneity in species richness can be primarily explained by the duration of evolution. Colville et al. [68]. In South Africa, 4813 plant species from Cape Floristic Province and 21 molecularly dated endemic clade were added to the simulation study, and the age and area hypothesis was tested taking into account known climates and topography going back 140,000 years. The regression model showed that long-term stability is the deciding factor in explaining species richness. In general, fossil analyses of both aquatic and terrestrial clades have shown higher rates of speciation (origination) in tropical areas [69, 70]. In the sea, the gradient is less steep because temperatures change less and organisms do not have to adapt to such large fluctuations as on land [71]. The gradient is quite conservative, dispersing events occurring in 4–5% of temperate regions. In most cases, dispersion occurs in other tropical regions. Duchêne and Cardillo [72] came to this conclusion in their study by integrating the phylogeny of 9000 bird species. Dispersal events toward temperate areas are generally no older than the Eocene-Oligocene Climate Transition. Rivadeneira et al. [73] used data from 328 marine mollusk species and 159 genera to explore the evolutionary processes that lead to the emergence of the current observed distribution. To do this, the fossil record, nestedness analysis, and projection matrix are used as complementary approaches. Geographical distribution was nested irrespective of the region of origin of genera, and according to the distribution dynamics model, dispersion events were common from temperate areas to the tropics, where extinction events were much rarer. In conclusion, despite the difficulty of distinguishing the signs of speciation and extinction in phylogenetic studies of diversification, all large-scale phylogenic research papers provide findings that are consistent with the tropical conservatism hypothesis and the area/energy/time hypothesis, namely that species have been concentrated in tropical bioregions to a greater extent than extratropical bioregions. The climate stability hypothesis complements the area/energy/time hypothesis in the sense that it is based on a similar mechanistic basis: the genetic variability of population and the geographical extent and, in return, these factors influence speciation and extinction rates. Building on the projections that a larger area and greater genetic variability can influence species formation and extinction, climate stability postulates that these factors are highly mediated by orbitally enforced range dynamics. Therefore, the time-integrated area of temperate climate regions is likely to fluctuate strongly during cold periods and increase the extinction rate. As the chances of extinction increase due to minimum temperatures, a cumulative time-integrated area may not be the best way to test the relationship between species richness and total area available for populations within a region. Instead, if contraction has resulted in extinction events, it is worth working with the minimum area when determining the species richness. Paleoclimatic changes during the Paleo and Neogene allowed or hindered species wandering between areas, as well as regional extinctions, leaving a mark on species and genera distribution. Few species and genera are disjunct between eastern North America and East Asia among temperate plants, suggesting past connectivity over the Bering land bridge as well as climate-driven extinctions from Europe and western North America [74]. Quaternary glacial-interglacial oscillations have left legacies in existing species ranges, according to a wide body of evidence. Ordonez and Svenning [75] examined the distribution patterns of Europe's contemporary vegetation using six independent lineages (Caryophyllales, Brassicales, Ranunculales, Saxifragales, Rosales, and Malpighiales). The Pleistocene climate, the former location and extent of refugium, the accessibility of areas after the Ice Age, and the contemporary environment play a major role in their current distribution. In concordance with a previous study, Costa et al. [76] concluded that climate stability played key role in Holocene biodiversity of the Amazonian-basin using 30 kyr pollen record and random forest classification. Decreases or fluctuations in temperature negatively correlate with species richness [77]. Deep-sea invertebrates are exception to the role because they do not exhibit equator-pole diversity gradient. Marine invertebrates can be much more diverse in deep waters, which are always roughly 4°C, than in shallow seas. But even in this case, stability is very important. So, constant temperature and access to resources have no variance over time. Deep sea is not productive at all, but resource availability is constant throughout the year [78]. A study of the last 500 million years shows that a steep gradient exists only in the last 30 million years due to cooling. During warm periods, the gradient was much shallower or it was turned around, and there were times when the temperate zone had maximum biodiversity. At that time, the climate was much more stable throughout the world, with few fluctuations [79]. High temperatures have direct positive effect on ectothermic animals, as physiological processes accelerate and evolutionary processes can be faster. Faster metabolism leads to shorter generational times, which, together with a higher mutation rate, leads to the formation of new species [80].

Due to the impact of temperature on fostering reproductive isolation, the evolutionary pace hypothesis suggests that speciation rate may be higher in warmer bioregions. Orton et al. [49] conducted a comprehensive study involving multiple phyla (Arthropoda, Chordata, Mollusca, Annelida, Echinodermata, and Cnidaria), COI (Cytochrome c Oxidase subunit I) sequences, and 8037 lineages, but only 51.6% of pairs exhibit higher mutation rate near the equator. For the remainder, the mutation rate was higher in the higher latitudes. This is the most comprehensive study in the literature that has occurred in recent years, and although the result has proved significant, the results are not entirely conclusive to prove evolutionary speed hypothesis.

3. Biotic factors

Organisms also interact with one another, and the effect on one another other also influences evolution. Andresen et al. [81] proved relationships are stronger and more diverse than in the temperate zone. Four different hypotheses attempt to explain species richness: Two involve the speciation (enemy-mediated habitat and the geographic mosaic of coevolution) and two relate to the existence and retention of diversity (competition causing finer niches and predation promoting coexistence). If access to resources differs in two different habitats, the plant will find it more difficult to regenerate the damage caused by herbivores and parasites in places where nutrients are less accessible. Therefore, they need to produce several compounds that keep animals away. However, due to protection, they grow more slowly and are less competitive with specimens living in nutrient-rich areas [82]. Because there are more enemies in productive environments in the tropics, this effect is also stronger. During the formation of the new species, it specializes in a particular habitat with growth protection optimization [83]. Different selection and speciation probabilities between interdepending mutualist or antagonistic species, in tropical climates, should be higher. Organisms are rarely killed by abiotic stress in tropical climates, with most of the selection processes in the tropics controlled by biotic factors. These factors evolve themselves and give an impetus to the evolution of the other species. In tropical environments, where seasonal differences are small, there are several interacting species (plants, herbivores, pathogens, predators, mutualists, etc.) and their relationships persist throughout the year. In cold climates, seasonality disrupts relations, slowing down co-evolution. In stable warm environments, co-evolution is faster and easier to develop mutualistic interrelationships [84]. The novel protection mechanism protects the plant from enemies, which enables the plant to increase its range, stimulating allopatric specificity. But if this enemy is able to adapt to the defense mechanism, then the range of the plant may shrink. Similarly, specialization may lead to speciation through the development of the host race or the increase in patches of geographical specialists that lead to differentiation. These processes will also increase over time and generate a positive feedback loop [85]. An example is the relationship between butterflies in the Nymphalidae family and solanaceae feed plants. The Solanaceae family separated from its sister groups 49–68 million years ago and then diversified (29–47 MYA). Around this time, the Ithomiini subfamily of the Nymphalidae family was also formed and diversified [86]. But there are also examples where this diversification effect has not been confirmed. According to Kaczvinsky and Hardy [87], the emergence of a new plant-predator relationship could increase the chances of extinction. The development of a connection with the new plant can be due to the significant fitness reduction of ancient feed plant. This can be caused by several factors, such as increasing competition for diminishing resources, the development of a new defense mechanism, the emergence of a new invasive species, and the emergence of new natural enemies. If a change in the feeding plant of an insect is caused by such an effect, it will cause a decrease in the population size and chance of speciation. With decreased overall performance of ancestral host and presumably minimal overall performance of new host, the growth rates and adequate sizes of herbivorous insect populations should shrink, as well as their geographic and climatic area.

Two other hypotheses explain that in areas where interrelationships between organisms are stronger, the width of the niche is smaller, in the sense that it uses habitat and resources. As competition intensifies, niches can split up, resulting in a finer niche for evolving new species. Thus, the existence of more species can be imagined within the community [22]. Harmáčková et al. [88] tested the hypothesis by using 298 Australian songbirds' (Passeriformes) data. It was expected that the species richness-specialization relationship is stronger in particularly species-rich communities, where annual precipitation is high and vegetation is complex. They also tested the extent of niche overlap. A positive relationship was found between specialization and species richness, but the direction and strength of relationship vary according to traits and area size. The specialization-species richness relationship is clear only in the forage stratum and has increased toward a smaller area only for habitat and diet. At the same time, local communities had a high overlap in habitat and diet. In a particularly species-rich community, no particularly strong link was found between species richness and specialization. However, they found a negative connection between specialization and overlap, meaning that species separate the ecological space on the basis of where food is found. These just weakly supported their expectations. The specialization in forage stratum has probably been significant in promoting species coexistence. On the other hand, while several species were habitat and diet specialists, high overlap in these traits did not rule out coexistence. The stronger predator (or herbivore) is able to drive selection in the antipredator traits in the prey (or plant). This is the basis of predator-victim specialization, the prey (or plant) can compete for a predator-free environment, leading to greater coexistence of the prey (or plant) species. Horst and Venable [89] studied the seed predation of rodents in the Sonora Desert. The rodents prefer the seeds of a certain species out of three different plants, which allows the two other plants to coexist. Regular predation on the more common species reduces interspecific competition between the three plants.

Phylogenetic studies of trees, birds, mammals, reptiles, and amphibians show that tropical regions are the source of biodiversity, composed of both ancient and recently evolved lineages. Moderate regions contain only a fraction of this. Over the past millions of years, the climate of tropical areas has been more stable, which has helped new species to evolve and reduced extinction rate. And niche conservatism has prevented lineages to wander to the temperate zone. It seems that the energy available is a secondary factor because deep sea life has become almost completely independent of it. But temperature and availability of resources are stable there in the long run. The time-integrated area is the most important controller of evolutionary processes. The seasonality of temperature and productivity affects the number of resources available on land, which determines the size of the populations. And lastly, speciation and coexistence are augmented by biotic interactions.

Author details

Levente Hufnagel[1]* and Ferenc Mics[2]

1 Research Institute of Multidisciplinary Ecotheology, John Wesley Theological College, Budapest, Hungary

2 Department of Environmental Security, John Wesley Theological College, Budapest, Hungary

*Address all correspondence to: leventehufnagel@gmail.com

IntechOpen

References

[1] Briggs JC. Ocean Biogeography. International Encyclopedia of Geography: People, the Earth, Environment and Technology. Hoboken, New Jersey, U.S.: John Wiley & Sons Ltd; 2016. pp. 1-5

[2] Cirtwill AR, Stouffer DB, Romanuk TN. Latitudinal gradients in biotic niche breadth vary across ecosystem types. Proceedings of the Royal Society B: Biological Sciences. 2015;**282**:20151589

[3] Maestri R, Patterson BD. Patterns of species richness and turnover for the south american rodent fauna. PLoS One. 2016;**11**(3):e0151895

[4] Kinlock NL, Prowant L, Herstoff EM, Foley CM, Akin-Fajiye M, Bender N, et al. Explaining global variation in the latitudinal diversity gradient: Meta-analysis confirms known patterns and uncovers new ones. Global Ecology and Biogeography. 2017;**27**(1):125-141

[5] Saupe EE, Myers CE, Townsend Peterson A, Soberón J, Singarayer J, Valdes P, et al. Spatio-temporal climate change contributes to latitudinal diversity gradients. Nature Ecology & Evolution. 2019;**3**:1419-1429

[6] Etienne RS, Cabral JS, Hagen O, Hartig F, Hurlbert Pellissier L, Pontarp M, et al. A minimal model for the latitudinal diversity gradient suggests a dominant role for ecological limits. The American Naturalist. 2019;**194**(5):122-133

[7] Fordyce JA, DeVries PJ. A tale of two communities: Neotropical butterfly assemblages show higher beta diversity in the canopy compared to the understory. Oecologia. 2016;**181**:235-243

[8] Willig MR, Presley SJ. Latitudinal gradients of biodiversity: Theory and empirical patterns. In: DellaSala DA, Goldstein MI, editors. The Encyclopedia of the Anthropocene. Oxford: Elsevier; 2018

[9] Hudson L, Newbold T, Contu S, Hill S, Lysenko I, De Palma A, et al. The database of the PREDICTS (projecting responses of ecological diversity in changing terrestrial systems) project. Ecology and Evolution. 2017;**7**:145-188

[10] Bernatchez L. On the maintenance of genetic variation and adaptation to environmental change: Considerations from population genomics in fishes. Journal of Fish Biology. 2016;**89**: 2519-2556

[11] Matuszewski S, Hermisson J, Kopp M. Catch me if you can: Adaptation from standing genetic variation to a moving phenotypic optimum. Genetics. 2015;**200**(4): 1255-1274

[12] Sollars ESA, Harper AL, Kelly LJ, Sambles CM, Ramirez-Gonzalez RH, Swarbreck D, et al. Genome sequence and genetic diversity of European ash trees. Nature. 2016;**541**:212-216

[13] Matz MV, Treml EA, Aglyamova GV, Bay LK. Potential and limits for rapid genetic adaptation to warming in a Great Barrier Reef coral. PLoS Genetics. 2018;**14**(4):e1007220

[14] Raffard A, Santoul F, Cucherousset J, Blanchet S. The community and ecosystem consequences of intraspecific diversity: A meta-analysis. Biological Reviews. 2019;**94**(2):648-661

[15] Siefert A, Violle C, Chalmandrier L, Albert CH, Taudiere A, Fajardo Aarssen LW, et al. A global meta-analysis of the relative extent of intraspecific trait variation in plant

communities. Ecology Letters. 2015;**18**(12):1406-1419

[16] Suo AN, Lin Y, Sun YG. Impact of sea reclamation on zoobentic community in adjacent sea area: A case study in Caofeidian, North China. Applied Ecology and Environmental Research. 2017;**15**(3):871-880

[17] Jahanbakhsh GM, Khorasani N, Morshedi J, Danehkar A, Naderi M. Factors influencing abundance and species richness of overwintered waterbirds in Parishan International Wetland in Iran. Applied Ecology and Environmental Research. 2017;**15**(4):1565-1579

[18] Fornal-Pieniak B, Ollik M, Schwerk A. Impact of surroundings landscape structure on formation of plant species in aforestrated manor parks. Applied Ecology and Environmental Research. 2018;**16**(5):6483-6497

[19] Rajamurugan J, Mohandass D, Campbell MC, Jayakrishnan P, Balachandran N, Shao S-C. Fragmentation causes woody plant composition decline in sacred grove patches in the Puducherry Region of Shoutheast. Applied Ecology and Environmental Research. 2021;**19**(3):1625-1643

[20] Zhang GX, Yuan XZ, Wang KH, Zhang MJ, Zhou LL, Zhang QY, et al. Biodiversity conservation in agricultural landscapes: An ecological opportunity for coal mining subsidence areas. Applied Ecology and Environmental Research. 2020;**18**(3):4283-4308

[21] Tufan-Çetin Ö. Determination of lichen diversity variations in habitat type of mediterranean maquis and arborescent matorral. Applied Ecology and Environmental Research. 2019;**17**(4):10173-10193

[22] Pianka ER. Latitudinal gradients in species diversity: A review of concepts. The American Naturalist. 1966;**100**(910):33-46

[23] Jablonski D, Huang S, Roy K, Valentine JW, Bronstein JL. Shaping the latitudinal diversity gradient: New perspectives from a synthesis of paleobiology and biogeography. The American Naturalist. 2017;**189**(1):1-12

[24] Schluter D, Pennell MW. Speciation gradients and the distribution of biodiversity. Nature. 2017;**546**:48-55

[25] Fine PVA. Ecological and evolutionary drivers of geographic variation in species diversity. Annual Review of Ecology, Evolution, and Systematics. 2015;**46**:369-392

[26] Sebastian-Azcona J, Harmann A, Hacke UG, Rweyongeza D. Survival, growth and cold hardiness tradeoffs in white spruce populations: Implications for assisted migration. Forest Ecology and Management. 2019;**433**: 544-552

[27] Bucher SF, Feiler R, Buchner O, Neuner G, Rosbakh S, Leiterer M, et al. Temporal and spatial trade-offs between resistance and performance traits in herbaceous plant species. Environmental and Experimental Botany. 2019;**157**:187-196

[28] Preisser W. Latitudinal gradients of parasite richness: A review and new insights from helminths of cricetid rodents. Ecography. 2019;**42**(7): 1315-1330

[29] Kolomiytsev N, Poddubnaya N. Temporal and spatial variability of environments drive the patterns of species richness along latitudinal, elevational, and depth gradients. Communications Biology. 2018;**63**(3):189-201

[30] Du C, Chen J, Jiang L, Qiao G. High correlation of species diversity patterns between specialist herbivorous insects

and their specific hosts. Journal of Biogeography. 2020;**00**:1-14

[31] Marin J, Hedges SB. Time best explains global variation in species richness of amphibians, birds and mammals. Journal of Biogeography. 2016;**43**(6):1069-1079

[32] Pyron RA, Costa GC, Patten MA, Burbrink FT. Phylogenetic niche conservatism and the evolutionary basis of ecological speciation. Biological Reviews. 2015;**90**(4):1248-1262

[33] Qian H, Ricklefs RE. Out of the tropical lowlands: Latitude versus elevation. Trends in Ecology & Evolution. 2016;**31**(10):738-741

[34] Schubert M, Grønvold L, Sandve SR, Hvidsten TR, Fjellheim S. Evolution of cold acclimation and its role in niche transition in the temperate grass subfamily pooideae. Plant Physiology. 2019;**180**:404-419

[35] Moss JA, Henriksson NL, Pakulski JD, Snyder RA, Jeffrey WH. Oceanic microplankton do not adhere to the latitudinal diversity gradient. Microbial Ecology. 2020;**79**:511-515

[36] Edwards EJ, Donoghue MJ. Is it easy to move and easy to evolve? Evolutionary accessibility and adaptation. Journal of Experimental Botany. 2013;**64**(13):4047-4052

[37] Patiño J, Weigelt P, Guilhaumon F, Kreft H, Triantis KA, Naranjo-Cigala A, et al. Differences in species–area relationships among the major lineages of land plants: A macroecological perspective. Global Ecology and Biogeography. 2013;**23**(11):1275-1283

[38] Castiglione S, Mondanaro A, Melchionna M, Serio C, Di Febbraro M, Carotenuto F, et al. Diversification rates and the evolution of species range size frequency distribution. Frontiers in Ecology and Evolution. 2017;**5**:147

[39] Polechová J, Barton NH. Limits to adaptation along environmental gradients. Proceedings of the National Academy of Sciences of the United States of America. 2015;**112**(20): 6401-6406

[40] Chichorro F, Juslén A, Cardoso P. A review of the relation between species traits and extinction risk. Biological Conservation. 2019;**237**:220-229

[41] Weeks AR, Stoklosa J, Hoffmann AA. Conservation of genetic uniqueness of populations may increase extinction likelihood of endangered species: The case of Australian mammals. Frontiers in Zoology. 2016;**13**:31

[42] Svenning J, Eiserhardt W, Normand S, Ordonez A, Sandel B. The influence of paleoclimate on present-day patterns in biodiversity and ecosystems. Annual Review of Ecology, Evolution, and Systematics. 2015;**46**(1): 551-572

[43] Anisha D, Akash G. Cenozoic era. In: Vonk J, Shackelford TK, editors. Encyclopedia of Animal Cognition and Behavior. Switzerland: Springer Nature; 2021

[44] Woolley SNC, Tittensor DP, Dunstan PK, Guillera-Arroita G, Lahoz-Monfort JJ, Wintle BA, et al. Deep-sea diversity patterns are shaped by energy availability. Nature. 2016;**533**(7603):393-393

[45] Belmaker J, Jetz W. Relative roles of ecological and energetic constraints, diversification rates and region history on global species richness gradients. Ecology Letters. 2015;**18**:563-571

[46] Armitage DW. Experimental evidence for a time-integrated effect of productivity on diversity. Ecology Letters. 2015;**18**(11):1216-1225

[47] Pocheville A. (2015): The ecological niche: History and Recent

Controversies. In: Heams T, Huneman P, Lecointre G, Silberstein M. (eds). Handbook of Evolutionary Thinking in the Sciences. Dordrecht: Springer; 2015

[48] Rabosky DL, Hurlbert AH, Price T. Species richness at continental scales is dominated by ecological limits. The American Naturalist. 2015;**185**(5): 572-583

[49] Orton MG, May JA, Ly W, Lee DJ, Adamowicz SJ. Is molecular evolution faster in the tropics? Heredity. 2019;**122**(5):513-524

[50] Dong Y, Wu N, Li F, Chen X, Zhang D, Zhang Y, et al. Influence of monsoonal water-energy dynamics on terrestrial mollusk species-diversity gradients in northern China. Science of the Total Environment. 2019; **676**:206-2014

[51] Xu X, Wang Z, Rahbek C, Sanders NJ, Fang J. Geographical variation in the importance of water and energy for oak diversity. Journal of Biogeography. 2015;**43**(2):279-288

[52] Pontarp M, Wiens JJ. The origin of species richness patterns along environmental gradients: Uniting explanations based on time, diversification rate and carrying capacity. Journal of Biogeography. 2016;**44**(4):722-735

[53] Gatti RC. A conceptual model of new hypothesis on the evolution of biodiversity. Biologia. 2016;**71**(3): 343-351

[54] Condamine FL, Rolland J, Morlon H. Assessing the causes of diversification slowdowns: Temperature-dependent and diversity-dependent models receive equivalent support. Ecology Letters. 2019;**22**: 1900-1912

[55] Pannetier T, Martinez C, Bunnefeld L, Etienne RS. Branching patterns in phylogenies cannot

distinguish diversity-dependent diversification from time-dependent diversification. Evolution. 2021;**75**(1):25-38

[56] Machac A, Graham CH. Regional diversity and diversification in mammals. The American Naturalist. 2017;**189**(1):1-13

[57] Zeng Y, Wiens JJ. Species interactions have predictable impacts on diversification. Ecology Letters. 2021;**24**(2):239-248

[58] Šímová I, Storch D. The enigma of terrestrial primary productivity: Measurements, models, scales and the diversity–productivity relationship. Ecography. 2017;**40**(2):239-252

[59] Storch D, Okie JG. The carrying capacity for species richness. Global Ecology and Biogeography. 2019;**28**: 1519-1532

[60] Storch D, Bohdalkova E, Okie J. He more-individuals hypothesis revisited: The role of community abundance in species richness regulation and the productivity–diversity relationship. Ecology Letters. 2018;**21**(6):920-937

[61] Mateo RG, Mokany K, Guisan A. Biodiversity models: What if unsaturation is the rule? Trends in Ecology & Evolution. 2017;**32**(8): 556-566

[62] Hopkins MJG. Are we close to knowing the plant diversity of the Amazon? Annals of the Brazilian Academy of Sciences. 2019;**91**(Suppl. 3): e20190396

[63] Piek M. Milankovitch cycles: Precession discovered and explained from Hipparchus to Newton [bachelor thesis]. Utrecht: University of Utrecht; 2015

[64] Simões M, Breitkreutz L, Alvarado M, Baca S, Cooper JC, Heins L, et al. The evolving theory

of evolutionary radiations. Trends in Ecology and Evolution. 2016;**31**(1):27-34

[65] Schluter D, Bronstein JL. Speciation, ecological opportunity, and latitude. The American Naturalist. 2016;**187**(1): 1-18

[66] Grace JB, Anderson TM, Seabloom EW, Borer ET, Adler PB, Harpole WS, et al. Integrative modelling reveals mechanisms linking productivity and plant species richness. Nature. 2016;**529**:390-393

[67] Marin J, Rapacciuolo G, Costa GC, Graham CH, Brooks TM, Young BE, et al. Evolutionary time drives global tetrapod diversity. Proceedings of the Royal Society B: Biological Sciences. 2018;**285**:20172378

[68] Colville JF, Beale CM, Forest Altwegg R, Huntley B, Cowling RM. Plant richness, turnover, and evolutionary diversity track gradients of stability and ecological opportunity in a megadiversity center. Proceedings of the National Academy of Sciences of the United States of America. 2020; **117**(33):20027-20037

[69] Cowman PF, Parravicini V, Kulbicki M, Floeter SR. The biogeography of tropical reef fishes: Endemism and provinciality through time. Biological Reviews. 2017;**92**(4): 2112-2130

[70] Tamma K, Ramakrishnan U. Higher speciation and lower extinction rates influence mammal diversity gradients in Asia. BMC Evolutionary Biology. 2015;**15**:11

[71] Yasuhara M, Danovaro R. Temperature impacts on deep-sea biodiversity. Biological Reviews. 2016;**91**(2):275-287

[72] Duchêne DA, Cardillo M. Phylogenetic patterns in the geographic distributions of birds support the tropical conservatism hypothesis. Global Ecology and Biogeography. 2015;**24**(11):1261-1268

[73] Rivadeneira MM, Alballay AH, Villafaña JA, Raimondi PT, Blanchette CA, Fenberg PB. Geographic patterns of diversification and the latitudinal gradient of richness of rocky intertidal gastropods: The 'into the tropical museum' hypothesis. Global Ecology and Biogeography. 2015;**24**: 1149-1158

[74] Xiang J-Y, Wen J, Peng H. Evolution of the eastern Asian–North American biogeographic disjunctions in ferns and lycophytes. Journal of Systematics and Evolution. 2015;**53**(1):2-32

[75] Ordonez A, Svenning J-C. Consistent role of quaternary climate change in shaping current plant functional diversity patterns across European plant orders. Scientific Reports. 2017;**7**:42988

[76] Costa GC, Hampe A, Ledru M-P, Martinez PA, Mazzochini GG, Shepard DB, et al. Biome stability in South America over the last 30 kyr: Inferences from long-term vegetation dynamics and habitat modelling. Global Ecology and Biogeography. 2018;**27**(3):285-297

[77] Rasconi S, Winter K, Kaintz MJ. Temperature increase and fluctuation induce phytoplankton biodiversity loss—Evidence from a multi-seasonal mesocosm experiment. Ecology and Evolution. 2017;**7**(9):2936-2946

[78] Chaudhary C. Global-scale distribution of marine species diversity: An analysis of latitudinal, longitudinal and depth gradients [PhD thesis]. New Zealand: University of Auckland; 2019

[79] Mannion PB, Upchurch P, Benson RBJ, Goswami A. The latitudinal biodiversity gradient through deep time. Trends in Ecology & Evolution. 2014;**29**:42-50

[80] Puurtinen M, Elo M, Jalasvuori M, Kahilainen A, Ketola T, Kotiaho JS, et al. Temperature-dependent mutational robustness can explain faster molecular evolution at warm temperatures, affecting speciation rate and global patterns of species diversity. Ecography. 2016;**39**(11):1025-1033

[81] Andresen E, Arroyo-Rodríguez V, Escobar F. Tropical biodiversity: The importance of biotic interactions for its origin, maintenance, function, and conservation. In: Dáttilo W, Rico-Gray V. Ecological Networks in the Tropics. Cham:Springer; 2018

[82] Jia S, Wang X, Yuan Z, Lin F, Ye J, Lin G, et al. Tree species traits affect which natural enemies drive the Janzen-Connell effect in a temperate forest. Nature Communications. 2020;**11**:286

[83] Viswanathan A, Ghazoul J, Lewis OT, Honwad G, Bagchi R. Effects of forest fragment area on interactions between plants and their natural enemies: Consequences for plant diversity at multiple spatial scales. Frontiers in Forests and Global Change. 2020;**2**:88

[84] Medeiros LP, Garcia G, Thompson JN, Guimaraes PR. The geographic mosaic of coevolution in mutualistic networks. Proceedings of the National Academy of Sciences of the United States of America. 2018;**115**(47):12017-12022

[85] Harmon LJ, Andreazzi CS, Débarre F, Drury J, Goldberg EE, Martins AB, et al. Detecting the macroevolutionary signal of species interactions. Journal of Evolutionary Biology. 2019;**32**(8):769-782

[86] Khyade VB, Gaikwad PM, Vare PR. Explanation of nymphalidae butterflies. International Academic Journal of Science and Engineering. 2018;**5**(4): 24-47

[87] Kaczvinsky C, Hardy NB. Do major host shifts spark diversification in butterflies? Ecology and Evolution. 2020;**0**(0):1-11

[88] Harmáčková L, Remešová E, Remeš V. Specialization and niche overlap across spatial scales: Revealing ecological factors shaping species richness and coexistence in Australian songbirds. Journal of Animal Ecology. 2019;**88**:1766-1776

[89] Horst JL, Venable DL. Frequency-dependent seed predation by rodents on Sonoran Desert winter annual plants. Ecology. 2018;**99**(1):196-203

Chapter 2

Biodiversity Conservation, Economic Growth and Sustainable Development

Richard E. Rice

Abstract

A growing economy has long been regarded as important for social and economic progress. And indeed, much of what we value in society is the product of economic growth. It is becoming increasingly clear, however, that growth cannot continue forever and that there is a price to pay for our failure to chart a more sustainable path. This chapter examines the conflict between our global obsession with growth and the conservation of biological diversity. The chapter begins with a discussion of what growth means and why it is the focus of global economic policy. We then review the connection between economic growth, sustainable development and the conservation of biological diversity and examine issues surrounding the quest for sustainable development, including how growth is measured and why there is a need to develop alternatives measures of growth and alternatives to a focus on perpetual growth. The chapter concludes with a discussion of the role that economic incentives can play in helping to catalyze necessary change and the importance of a commitment to cost-effectiveness in the choice of policies to promote conservation action.

Keywords: Biological diversity, economic development, Sustainability, GDP, Genuine Progress Indicator, conservation agreements, carbon taxes

1. Introduction

Since its introduction during World War II most countries have come to view gross domestic product, or GDP, as their main measure of economic progress. Growth in GDP is widely seen as essential for advancing human welfare, even as the implications of this growth ever more clearly present us with existential threats, including a rapidly changing climate and dire impacts on biodiversity. With record growth have come record droughts and heatwaves. The last seven years, in fact, have been the warmest since records began in 1880 and last year, 2020, tied 2016 as the warmest year ever [1]. Wildfires across the planet are growing larger and more frequent and ever more evidence accumulates that ecosystems around the globe are collapsing [2–10].

Each day's news it seems underscores the fact that there is a price to pay for our global obsession with growth and limits to what the biosphere can provide to an ever-larger global economy. As a result, the pressure for growth is increasingly being met with calls for greater sustainability. How these two things can be reconciled may be the most urgent and important challenge of our time.

This chapter will summarize the debate over the limits to economic growth beginning with a discussion of how growth is defined and why it is the focus of national economic policy. We will then review the connection between economic growth, sustainable development, and the conservation of biodiversity and examine issues surrounding the quest for sustainable development, including alternative measures of growth and alternatives to a focus on perpetual growth. We will end the chapter with a discussion of policies to help move the world onto a safer, saner trajectory focusing on the role that economic incentives can play in catalyzing necessary change and the importance of a commitment to cost-effectiveness in the design of policies to promote conservation action.

2. What growth means

The standard definition of economic growth is a sustained increase in a nation's real (inflation adjusted) gross domestic product (GDP). GDP is the monetary value of all goods and services produced in a country each year. In recent years, real GDP growth in the U.S. has averaged around 2% which means that the economy doubles in size every 36 years [11].

2.1 Why grow?

Proponents of economic growth focus on its many benefits, including higher standards of living and the ability to devote more resources to things like health care and education. Increases in sanitation, nutrition, and longevity have all been possible due to economic growth. Since 1800, life expectancy has grown from less than 30 years to more than 70 with eradication of childhood disease and improvements in medicine and nutrition [12]. Vast changes in material abundance have also been possible due to economic growth allowing many the things that only the wealthy could aspire to in the past.

Though something we now take for granted economic growth is a very recent phenomenon. Widespread economic prosperity (as measured by GDP per capita) has only been achieved in the past couple hundred years and as shown in **Figure 1**, has only really taken off in the past 50 years [13].

The incidence of extreme poverty over this period has fallen dramatically, in rich countries and poor alike [14]. Since 1990 alone the number of people living in extreme poverty has fallen by more than 1 billion [15]. The reasons for this reduction are many but one essential element has been the increase in crop yields achieved due to massive public investments in modern agricultural research. According to IFPRI [16], the case of English wheat is typical. Whereas it took nearly a millennium for yields to go from 0.5 to 2.0 metric tons per hectare it took only 40 years to rise from 2.0 to 6.0 metric tons per hectare. Yield increases such as these for wheat, rice and other crops have led to unprecedented levels of food security for many developing countries, despite large and continuing increases in population [16].

2.2 The downsides to growth

Given its many benefits, it is little wonder that economic growth is a focus of global economic policy. Growth, however, has its costs. Environmental destruction and impacts on biodiversity are perhaps the most obvious, but there are also

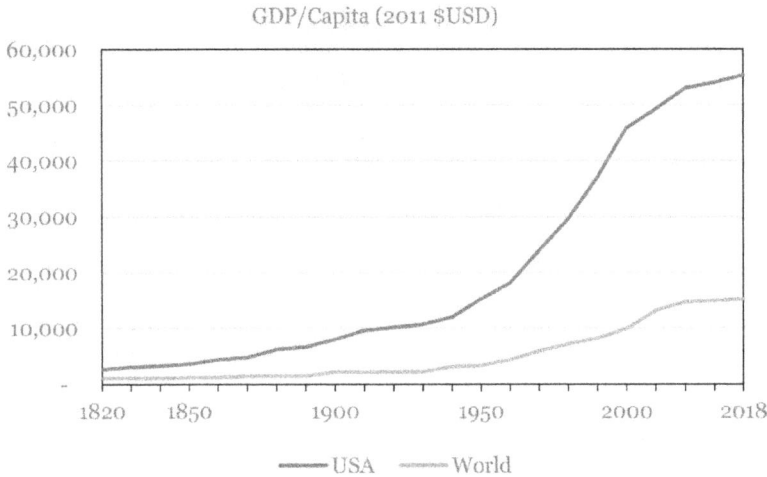

Figure 1.
The history of Economic growth: GDP/capita, 1820–2018 [13].

conflicts between economic growth and national security and international stability, and ultimately, economic sustainability itself.

Growing economies consume natural resources and produce wastes. This results in habitat loss, air and water pollution, climate disruption, and other environmental threats, threats which are becoming more apparent as economic activity encounters more and more limits. The depletion of groundwater and ocean fisheries are examples as are shortages of fresh water, and the global spread of toxic compounds such as mercury, chlorofluorocarbons, and greenhouse gases.

These conflicts are in part the result of the inescapable impact of an ever-growing human population. They are, however, exacerbated by market failures, including externalities and open-access resources, and in the case of biodiversity, the lack of markets altogether.

Externalities are the side-effects of commercial activities that impact third parties and are not reflected in the costs of production, and for this reason are "external" to the decision-making of both producers and consumers. Pollution from a factory is a negative externality. Intertemporal externalities (e.g., from climate change) impose costs on those in the future that are external to current generations. Externalities of all sorts undercut the ability of markets to produce sustainable outcomes.

Resources that are open to all without restriction, such as ocean fisheries, also invite unsustainable outcomes as is evidenced by the currently depleted state of the world's open-access fisheries.

Biodiversity suffers from a third market failure, the fact that it is generally not traded in formal markets. Though the popular conception of overexploitation is of resources plundered by the forces of markets, the absence of a market can be equally problematic. Things with no price end up being treated as if they have no value. Such is the fate of endangered species, tropical rainforests, coral reefs, and indeed much of wild nature.

Environmental impacts, of course, are not unconnected to society at large. Things like climate change and the extinction crisis have economic impacts and these in turn can threaten national security and international stability. Such threats are often made worse by inequality. Not everyone benefits equally from growth and some have arguably not benefitted at all. The problem of growing inequality is certainly an issue in the U.S. where the nation's top 10 percent now average more income than the bottom 90 percent [17]. But it is also clearly a problem globally. Sub-Saharan Africa is a case in point (see also, **Figure 1**). Although the poverty rate there has fallen in percentage terms since 1990, it has not fallen fast enough to keep pace with population growth [18]. As a result, the number of poor in that region continues to rise and now accounts for nearly two thirds of the world's total population in extreme poverty [18].

Climate change, resource scarcity, and environmental degradation generally are certain to accentuate such inequalities in the future with unavoidable impacts on social unrest, national security, and international stability. The national security implications of these issues were starkly presented in a recent report commissioned by the U.S. Army [19]. According to the study, America could face a grim series of events triggered by climate change involving drought, disease, failure of the country's power grid and a threat to the integrity of the military itself, all within the next two decades. The report also projects that sea level rise in the future is likely to "displace tens (if not hundreds) of millions of people, creating massive, enduring instability" and the potential for costly regional conflicts [19]. The report cites in particular the role that drought has played in sparking the civil war in Syria and the potential for tensions stemming from sea level rise and large-scale human displacement in Bangladesh.

All of the above issues have clear implications for economic sustainability – a healthy environment and international stability, after all, are the foundations for a healthy economy. We need healthy soils for agriculture, healthy oceans for fisheries, clean air and water and a stable political environment for international trade, all of which are threatened by unrestrained growth [20].

3. The quest for sustainable development

Increasing awareness of the limitations of growth has led to much discussion of sustainable development. This concept is most commonly associated with a report published by the World Commission on Environment and Development in 1987. In that report sustainable development is defined as "development that meets the needs of the present without compromising the ability of future generations to meet their own needs" [21]. Since the publication of this report, the idea of sustainable development has gained a solid footing in the popular imagination. An important landmark in this regard is the signing of the so-called Rio Declaration at the Earth Summit in 1992 in which 192 nations committed themselves to a detailed agenda for sustainable growth and development [22].

Despite its popularity, the precise meaning of sustainable development is somewhat elusive. From an economic perspective a simple definition might be that growth should proceed so long as the marginal benefits exceed the marginal costs (**Figure 2**). Marginal cost is the cost of a small increase in an activity and marginal benefit is the additional benefit from that increase. **Figure 2** shows the marginal costs and benefits of growth in GDP. Since the benefits tend to decline and the costs to rise with additional GDP growth, the sweet spot is to grow until the marginal costs are exactly equal to the marginal benefits. Any increase in GDP up to this point is "economic growth" whereas growth in GDP past this point, where costs rise above benefits is uneconomic [20].

3.1 The problems with GDP

Such definitions are all well and good, but problems arise in discerning when and where costs begin to exceed benefits. This, in turn, is made more difficult by

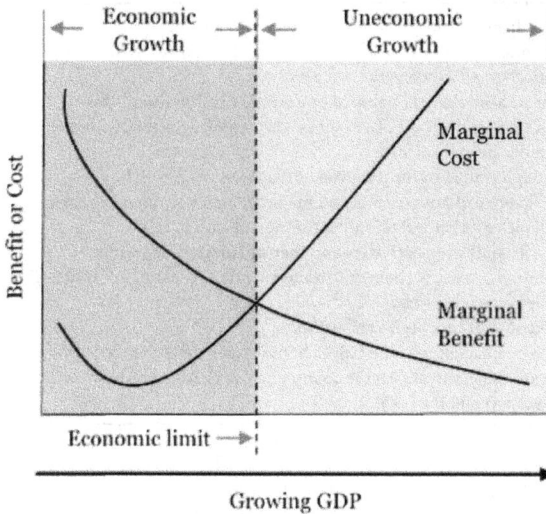

Figure 2.
Economic and uneconomic growth in GDP [20].

the way in which we measure growth. Ironically, GDP, our global standard measure of growth, was never intended as a measure of costs and benefits. Instead, it is simply a gross tally of market output with no distinction made between output that adds to well-being and output that diminishes it. Instead of separating costs from benefits GDP assumes that all monetary transactions by definition add to social welfare [23].

GDP also excludes everything that happens outside formal markets and therefore ignores many things that clearly benefit society such as volunteer work and unpaid work in households like childcare and elder care. Much of the value of environmental services is ignored as well.

As shown in **Box 1**, this method of accounting leads to some very counterintuitive results, including the fact that GDP increases with polluting activities and then again with clean-ups, crime and natural disasters are treated as economic gain, and the depletion of natural capital is treated as income [23].

The shortcomings of GDP are particularly significant with regard to biodiversity. As shown in **Box 2**, biodiversity underpins virtually all economic activity. Yet, it is not explicitly accounted for anywhere in GDP. In many cases, biodiversity is an unvalued input (e.g., crop and livestock genetics) into an output (food) whose value is counted in GDP. And while the connection between the two is clear in a general sense, the impact of added growth on the unvalued input is not. Worse, to the extent that further growth depletes the biodiversity we depend on it is counted as adding to national income. And since the benefits of avoiding the depletion of biodiversity often accrue to others (either in full or in part) there is little incentive for individuals or governments to invest in its conservation.

3.2 Moving beyond GDP

Faced with the obvious limitations of GDP, many countries are now looking for alternative ways of measuring social and economic health, including adjustments to measures like GDP and the development of alternative indicators.

GDP treats crime, divorce, and natural disasters as economic gain.

GDP counts all monetary transactions as positive. So, crime, divorce, and natural disasters, like fires and hurricanes, are all counted as economic progress.

GDP ignores the non-market economy of households and communities.

GDP ignores all activities that take place outside the market economy, including volunteer and home-based work such as childcare and elder care.

GDP treats the depletion of natural capital as income.

GDP treats the depletion of both natural and man-made capital as income rather than depreciation. So the more a country depletes its natural resources the more it adds to GDP.

GDP increases with polluting activities and then again with clean-ups.

GDP counts pollution as a double benefit to society by first including the economic activity that leads to pollution and then the cost of clean-ups.

GDP takes no account of income distribution.

GDP ignores income inequality. In the U.S. GDP has grown more than seven-fold since 1980 [24]. GDP presents this growth as a benefit to all, yet the country's three richest men now own more wealth than the bottom half of the country combined [25].

Box 1.
What's wrong with GDP? [23].

3.2.1 GDP adjustments

A basic problem with GDP and other conventional measures is that they are measures of output, not welfare. A true measure of welfare would rise when societies are better off and decline when they are worse off [26]. One of the limitations of GDP as a welfare indicator is that it does not take account of the depletion of either natural or man-made capital. As a result, spending to replace worn-out machinery is treated as income even though it adds nothing to the existing stock of machinery. Similarly, consumption and pollution that depletes society's store of natural capital is also incorrectly treated as income.

Food Security and Global Nutrition – Food production depends on biodiversity for plant and animal varieties, pollination, pest control, and disease regulation [27]. Indigenous produce adapted to local conditions in countries around the world serve as a basis for improved plant varieties and as a buffer against a changing climate [28, 29].

Disease Regulation – Lowered biodiversity and habitat fragmentation can lead to increased disease transmission and higher healthcare costs [30, 31]. Medicinal plants and manufactured pharmaceuticals rely on biodiversity. The diversity of plants and animals is an essential source of molecular compounds needed for future drug discovery [32].

Business and Livelihoods – More than half the world's GDP is moderately or highly dependent on nature, including nature-based tourism and recreational hunting and fishing [28, 33]. Fisheries, forestry and agriculture provide trillions of dollars annually in economic activity [34].

Protection and Replenishment – Biodiverse ecosystems provide natural buffers against storms and floods, water purification, soil formation and organic waste disposal [28]. Biodiversity underpins forests, grasslands, and agricultural systems essential for carbon storage and climate regulation [28].

Box 2.
Biodiversity underpins Economic activity, human health and wellbeing.

The former limitation can be addressed by simply subtracting an estimate of capital depreciation from GDP. This is now done as a matter of course in many countries, including the U.S. in what is called net domestic product (NDP) [35]. Adjusting for GDP's treatment of natural capital, however, is more complicated since there are uncertainties about precisely which cost items to deduct from GDP as well as how these items should be valued [36].

Nevertheless, in an effort to redress this shortcoming, economists have developed an alternative measure called the genuine progress indicator (GPI) which subtracts the value of natural capital used in production as well as the costs of negative externalities from GDP [37].

GPI also attempts to address other limitations of GDP by broadening the conventional accounting framework to include the benefits of volunteering and household labor as well as the impact of a variety of other factors, including crime, health care, income distribution, and leisure [37]. In effect, the GPI aims to serve as an indicator of sustainable welfare by focusing on the value of two basic things: activities that actually make us better off and those that are likely to be sustainable over the long term [37, 38].

Not surprisingly, GPI tells a rather different story than GDP of the recent history of economic growth. In an exhaustive study of the difference between the two indicators Kubiszewski, et al. [39] looked at 17 countries for which GPI data are available over the period 1950–2005. As shown in **Figure 3**, whereas GDP/capita rises continuously over this period, GPI/capita levels off in the late 1970s and begins to decrease slightly thereafter.

3.2.2 Alternative indices

Despite the theoretical appeal of the GPI, it too has limitations. Uncertainties about what costs and benefits to include and how they are valued tend to make these kinds on indices ill-defined. There are also unavoidable problems with trying to summarize how well a society or economy is doing using a single number.

These issues have given rise to specialized indices (e.g., of ecological health or happiness) as well a dash-board approach involving selected indicators that allow societies to better track the things they really aspire to.

One specialized index (the Living Planet Index) measures the state of global biodiversity based on population trends of vertebrate species from around the world. As shown in **Figure 4**, the most recent index shows an average 68% decline in the abundance of 4,392 mammal, bird, fish, reptile, and amphibian species from 1970 to 2016 [40]. Some groups are doing much worse. Freshwater populations have declined by an average of 84%, with regional declines as high as 94% (in Latin America). These startling reductions underscore the extent to which GDP as a standalone indicator is masking the impacts of economic growth.

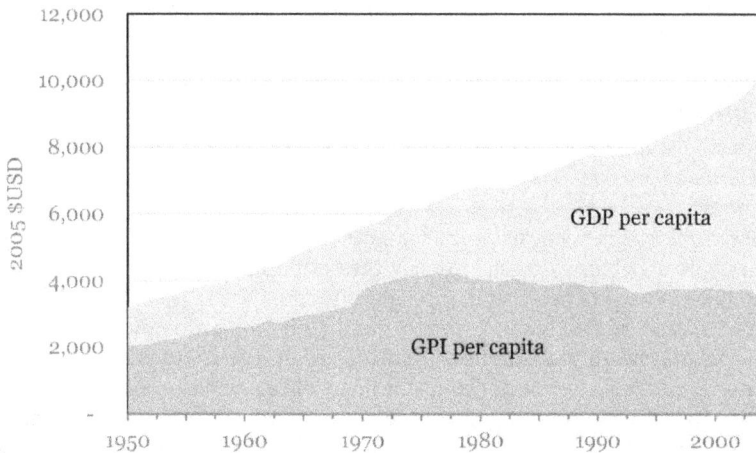

Figure 3.
GDP vs. GPI (genuine Progress indicator), 1950–2005 [39].

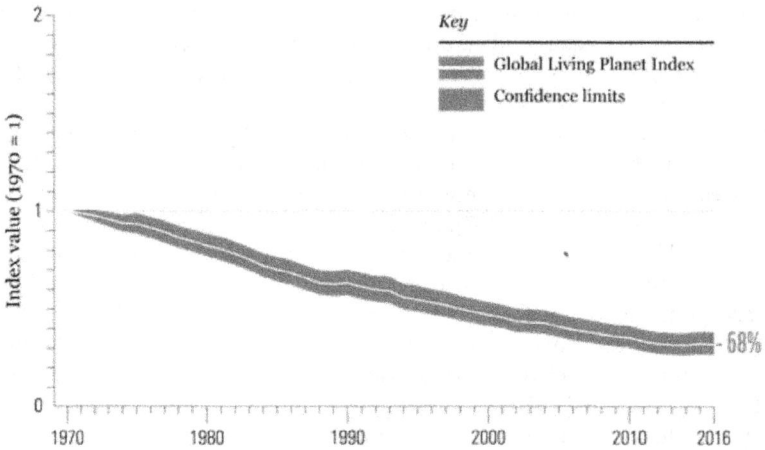

Figure 4.
The global living planet index (LPI) shows a 68% average decline between 1970 and 2016 [40].

An alternative to using a single index is the so-called dash-board approach, involving what are sometimes called sustainable development indicators. This approach seeks to go beyond measuring simply material wealth to focus on a broad range of indicators of the quality of life and environmental health.

One example of this approach is the Better Life Initiative [41] developed by the Organization for Economic Cooperation and Development (OECD), a group of 37 mostly rich countries. This initiative recommends 11 indicators that the OECD suggests as essential to well-being in terms of material living conditions (housing, income, jobs) and quality of life (community, education, environment, governance, health, life satisfaction, safety and work-life balance) [http://www.oecdbetter-lifeindex.org/#/45555545544].

At present, these indicators – which have been developed for all 37 OECD member countries – reflect only current well-being but in the future the organiza-tion expects to complement these with indicators describing the sustainability of well-being over time.

3.2.3 Concepts over numbers

A common shortcoming of all the above indicators is complexity. One reason for the power of GDP, despite its flaws, is simplicity. Up is good, down is bad, and even though a single, modified index like the GPI shares in this advantage, its usefulness as a measure of progress (or peril) is much diminished if it is unlikely to be accepted as a standard.

In response to this dilemma, some have opted for advancing concepts rather than numbers to help inspire and guide in the development of policies that will ultimately be needed to move us in the right direction. Two ideas worth mentioning in this regard are the steady state economy and doughnut economics.

The idea of a steady state economy is most closely associated with the work of economist Herman Daly, one of the co-founders of the journal *Ecological Economics*. According to Daly, a steady state economy seeks to respect the bounds of sustain-ability by keeping GDP and resource use stable [42]. As measured by GDP, an econ-omy is either growing, stable or in recession. Since neither economic growth nor recession is sustainable, a steady state economy is the only sustainable prospect and is therefore the "only appropriate policy goal for the sake of sustainability" [42].

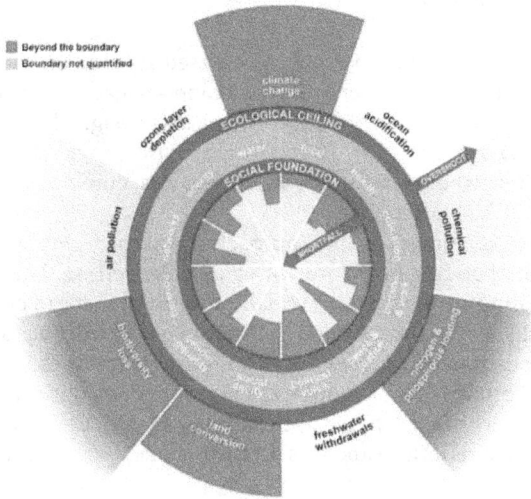

Figure 5.
The doughnut of social and planetary boundaries [image credit: Kate Raworth and Christian Guthier] [44].

Proponents of the steady state emphasize that it should not be confused with economic stagnation which, they say, is the result of a failed growth economy whereas a steady state economy seeks to balance the lack of traditional growth with efforts to distribute wealth so as to broaden economic security [43].

Doughnut economics, the creation of economist Kate Raworth, is in many ways a popularized version of Daly's steady state economy. Both authors reject the idea that perpetual growth is a viable option and instead call for maximizing social welfare within the physical and ecological limits of the planet. According to Raworth, the goal of economic activity should be to "meet the needs of all" while respecting planetary boundaries [44]. Raworth uses a doughnut, i.e., a disc with a hole in the middle, as her visual framework in which the inner ring represents society's social foundation and the outer ring its environmental ceiling (**Figure 5**). Between the two is what Raworth calls an "environmentally safe and socially just space in which humanity can thrive" [44].

4. Policies to take us there

The above discussion of how we define and measure sustainability, of course, begs the question of how we get from here to there. Clearly, a part of the answer lies in the measures and definitions themselves. We cannot correct problems if our measures conceal them, and we will never achieve sustainability if we do not define it as an explicit objective.

Nevertheless, this still leaves the difficult work of developing policies to help promote more sustainable outcomes. Experience and the existence of market failures suggests that we cannot leave solutions to the market alone. That said, it would be a mistake to underate the potential for productively using market forces in our search for solutions. Policies based on economic incentives in particular offer an extremely powerful and effective set of options.

Two examples in areas that matter to biodiversity are conservation agreements and carbon pricing. Both illustrate how incentive-based policies can help provide simple, cost-effective, and scalable solutions to environmental problems.

4.1 Conservation agreements

Conservation agreements are performance-based agreements in which resource owners commit to a concrete conservation outcome – usually the protection of a particular habitat or species – in exchange for benefits designed to give them an ongoing incentive to conserve [45]. The type of benefits provided vary but can include technical assistance, support for social services, employment in resource protection, or direct cash payments.

One of the great advantages of this approach is that the terms of agreements are flexible and can therefore be tailored to a particular setting. This flexibility makes conservation agreements a very scalable approach that can be implemented on private and indigenous lands outside traditional protected areas as well as on lands managed by national governments. In addition, whereas the creation of a traditional park or protected area requires a long, complex political process, conservation agreements, as a market-based approach, make park creation more akin to a standard business transaction, and this, in turn, makes park creation much more rapid and efficient.

Since conservation agreements are a voluntary approach that addresses the underlying costs of conservation they are more politically acceptable than forced buyouts or eminent domain and are also often less expensive than other approaches since they focus on opportunity cost which in many cases is extremely low, particularly in developing countries [46].

Conservation agreements were first piloted in 2001 in the context of a timber concession in Guyana [45]. Since then, they have been implemented in a wide variety of settings in roughly 20 countries around the world [47]. Examples include agreements focused on particular species as well ecosystems such as coral reefs, mangroves, and in the Solomon Islands, the largest uninhabited island in the South Pacific [47, 48].

4.2 Pricing carbon

Carbon pricing is another example of an incentive-based policy that relates to biodiversity. While this approach does not target biodiversity directly, it is perhaps the most important single policy affecting all life on Earth. When it comes to conservation, and so much else, unless we effectively tackle climate change very little else will matter.

Although there are many ways of putting a price on carbon, by far the simplest and most effective is a tax imposed on fuel suppliers (e.g., oil and gas producers). Once taxed, fuel suppliers raise their prices and in this way the higher prices ripple through the whole economy. There is no way to evade the tax and there is nothing to monitor or enforce (other than whether energy producers pay their taxes). Across the economy the cost of energy-intensive goods and services would rise giving both businesses and consumers an incentive to conserve.

One of the many advantages of a carbon tax is that it ensures that emission reductions are achieved at least cost to society. The reason is that unlike regulations that require everyone to adopt a particular technology or reduce their emissions by a certain amount, carbon taxes allow for the fact that some entities can reduce their emissions at a lower cost than others. This flexibility offers the opportunity for substantial cost savings.

Regulations alone, for example, can be twice as expensive as a carbon tax per ton of carbon abated while reducing far fewer emissions [49]. Similarly, subsidies (e.g., for electric vehicles) are unavoidably wasteful since they cannot target those who will only be motivated to buy because of the subsidy. If a tax credit of $7,500 convinces only one in four people to buy a hybrid electric vehicle, for example, the effective cost of the incentive is four times the subsidy or $30,000 – more than the price of many plug-in hybrids [50]. Such subsidies also tend to disproportionately benefit high-income households and while hybrids themselves emit less carbon than conventional cars, if the source of power used to charge them comes from coal they will raise carbon emissions rather than reduce them [51].

In addition to being less expensive, carbon taxes have several other important advantages. To begin, the cost of the tax is clearly known ahead of time. If the cost varies, as is true with cap and trade – the program used in several U.S. states – it makes it difficult for business (and consumers) to plan and therefore undercuts incentives to make long-term investments in efficiency.

Other options for pricing carbon are also more administratively burdensome and less transparent and often address only a subset of emissions. Cap and trade, for example, typically covers only electric utilities, which in the U.S. leaves out nearly three-quarters of total carbon emissions [52].

Most carbon tax proposals also now involve offsetting rebates so they do not disadvantage the poor who spend a larger percentage of their income on energy. Many proposals, in fact, would leave the majority of households better off with the tax than without it. In effect, such a "tax" would pay people for doing the right thing.

An important adjunct to a carbon tax is a UN program called REDD – Reducing Deforestation and Forest Degradation. REDD is a global effort designed to break with historic trends of increasing deforestation and greenhouse gas emissions by offering countries a financial incentive for forest conservation [53]. Since deforestation is the second largest anthropogenic source carbon emissions any realistic plan for addressing climate change must include efforts to halt the loss of tropical forests [54].

REDD takes advantage of the fact that reducing emissions anywhere on the globe has the same beneficial impact on slowing climate change. Reducing emissions through REDD therefore offers a means for offsetting emissions of industries that have no other option for meeting their climate commitments. For this reason, airlines around the world who have committed to being net-zero emitters in coming decades are expected to be major future funders of forest conservation through REDD [55].

Happily, protecting tropical forests is one of the least cost ways of reducing carbon emissions [56, 57]. REDD therefore has the potential for simultaneously reducing the cost of fighting climate change while providing a powerful incentive for protecting biodiversity.

4.3 A lack of environmental support

Given their advantages for conservation one might well expect that the three policies discussed above would be popular with environmentalists. In fact, all three policies have faced significant environmental opposition. Conservation agreements have received a great deal of favorable media attention but apart from modest investments by the organization that first developed them, they have largely been ignored by the international conservation community. This is in part a reflection of the fact that "paying for conservation" is regarded by many as a foreign concept, or worse, a dangerous precedent that "commodifies" nature and risks making all conservation efforts more expensive.

But it also reflects an important underlying incentive that shapes the conservation establishment. After years of strong popular support, the budgets and staff of all the major international conservation organizations have grown to the point where conservation has become an extremely expensive undertaking, one that depends critically on continued success in fundraising. And that, in turn makes for resistance to changes in tactics that would funnel money away from existing staff (even to laudable objectives like providing resource owners with an ongoing incentive to conserve). In the language of economics, the opportunity cost of supporting this kind of incentive-based conservation is the funding not going to current operations.

Carbon taxes have suffered from a similar lack of support. Part of the problem in this case is that taxes in general are an unpopular approach. But they have also suffered from competing agendas and a basic lack of understanding as illustrated by the fate two carbon tax bills in the U.S. state of Washington. The first was a revenue neutral bill that included tax cuts and rebates to offset the impact of higher prices from the carbon tax. This bill was defeated by an unusual coalition of oil interests and environmentalists. The later felt that the money collected by the government should be used to offset the impact of the tax on the poor (even though that is exactly what the rebates would have done) and to fund investments affecting climate, communities, and racial equity [58].

To accommodate these concerns, the second bill included no offsetting rebates and instead called for using the tax revenue to support a dedicated fund focused on the environment and social justice. In addition, the bill called for reducing the carbon tax by half to lessen its impact on prices. In effect, these changes made the revised bill both more regressive and less effective in reducing carbon emissions. Despite these "improvements", this bill was also defeated, this time by voters who objected to the added tax and the fact that it was being used to fund what the Seattle Times called a grab bag of "special interest payouts" [59].

The UN REDD program has also faced environmental objections, in this case based on concerns over the long-term security of emission reductions in developing countries and the fact that offsets allow polluters to avoid reducing their own emissions by paying for cheaper emission reductions elsewhere [60].

5. Summary and conclusion

The past two centuries of economic growth have provided the world with many benefits. Our lives are longer and healthier with more leisure and shorter workweeks. Childhood diseases that afflicted our parents are largely a thing of the past. The creative explosion of the last few decades has yielded advances in medicine, the arts, technology and more. All these things are the benefits of economic growth.

There are, however, downsides to economic growth that put our past progress and the future of life in jeopardy. Although global economic policy is still strongly wedded to growth in GDP there is increasing recognition that this is not a sustainable situation. Blindly promoting ever more growth without seeking to address market failures and impacts on the environment is clearly a prescription for trouble. The question is how to moderate these impacts while still maintaining a focus on advancing economic security and the quality of life.

Part of the answer to this question is in developing better indicators of how economic activity is affecting the things we care about. Having a global standard measure like GDP that ignores the value of nature and counts both pollution and clean up as progress is certain to steer us in the wrong direction. Dethroning GDP and work on replacements are worthy endeavors. Measures of impact, though, even at their best, are better at informing us of the need for change than in incentivizing specific changes. They still leave us with the hard work of developing appropriate policies for the future.

How we proceed in this regard will make a difference. Unconstrained markets are not likely to produce a happy ending, but this does not mean that we should ignore the potential for using markets and incentives in our search for solutions. The same forces that are driving us in the wrong direction can be harnessed and channeled in directions that will greatly enhance the potential for sustainable outcomes.

This is particularly true in the case of policies designed to address threats to biodiversity. Indeed, in the case of two important policies, carbon taxes and conservation agreements, ignoring this potential is likely to come at a price. Compared to a carbon tax, standards and subsidies could double the cost of dealing with climate change and rejecting the use of incentives in conservation agreements and REDD could jeopardize whether forests are saved at all.

The good news is that we have some extremely simple and powerful tools at our disposal. A single, small change in the tax code can reorient the entire economy away from carbon. And conservation agreements and REDD can be flexibly implemented almost everywhere they are needed. While funding these efforts will not be inexpensive there is ample global willingness and ability to pay for conservation and no shortage of those in a position to conserve who are willing to accept payment.

The challenges are great, but many of the tools needed to address them are at hand. We need only choose to put them to use.

Author details

Richard E. Rice
University of Maryland Global Campus, Silver Spring, MD, USA

*Address all correspondence to: richarderice@gmail.com

IntechOpen

References

[1] NASA. 2020 Tied for Warmest Year on Record, NASA Analysis Shows [Internet]. 2020. [cited 2021 Jun 6]. https://www.nasa.gov/ press-release/2020-tied-for-warmest-year-on-record-nasa-analysis-shows

[2] Patel, Six Trends to Know about Fire Season in the Western U.S. [Internet]. 2019. [cited 2021 Jun 6]. https://climate. nasa.gov/blog/2830/six-trends-to-know-about-fire-season-in-the-western-us/

[3] Gray, E. Satellite Data Record Shows Climate Change's Impact on Fires [Internet]. 2019. [cited 2021 Jun 6]. https://climate.nasa.gov/news/2912/ satellite-data-record-shows-climate-changes-impact-on-fires/

[4] Filbee-Dexter, K., Wernberg, T. Rise of turfs: A new battlefront for globally declining kelp forests, BioScience [Internet], 2018 Feb [cited 2021 Jun 29];(68)2:64-76. Available from: https:// doi.org/10.1093/biosci/bix147

[5] IUCN. IUCN Red List of Ecosystems [Internet]. 2021. [cited 2021 Jun 6]. https://iucnrle.org/resources/ published-assessments/

[6] Kareiva, P, Carranza, V. Existential risk due to ecosystem collapse: Nature strikes back. Science Direct [Internet]. 2018 [cited 2021 Jun 29];(102):39-50, Available from: https://www. sciencedirect.com/science/article/pii/ S0016328717301726 https://doi. org/10.1016/j.futures.2018.01.001 ISSN 0016-3287

[7] Perry, C, Murphy, G, Kench, P, et al. Caribbean-wide decline in carbonate production threatens coral reef growth. Nat Commun [Internet]. 2013 [cited 2021 Jun 29];(4):1402, Available from: https://doi.org/10.1038/ncomms2409

[8] Seibold, S, Gossner, M, Simons, N, et al. Arthropod decline in grasslands and forests is associated with landscape-level drivers. Nature [Internet]. 2019 [cited 2021 Jun 29];(574):671-674. Available from: https://doi.org/10.1038/ s41586-019-1684-3

[9] Stanke, H, Finley, A, Domke, G., et al. Over half of western United States' most abundant tree species in decline. Nat Commun [Internet]. 2020 [cited 2021 Jun 29];(12):451. Available from: https:// doi.org/10.1038/s41467-020-20678-z

[10] Swiss Re Institute. Biodiversity and ecosystem services: A business case for re/insurance [Internet]. Zurich. Swiss Re Institute. 2020. [cited 2021 Jun 29]. Available from: https://www.swissre. com/institute/research/topics-and-risk-dialogues/climate-and-natural-catastrophe-risk/expertise-publication-biodiversity-and-ecosystems-services. html

[11] Trading Economics. United States GDP Annual Growth Rate. 2021. [cited 2021 Jun 6]. https:// tradingeconomics.com/united-states/ gdp-growth-annual

[12] Roser, M, Ortiz-Ospina, E., Ritchie, H. Life Expectancy [Internet]. Our World in Data. 2019. [cited 2021 Jun 6]. https:// ourworldindata.org/life-expectancy

[13] Boldt, J, Luiten van Zanden, J. Madison style estimates of the evolution of the world economy. A new 2020 update [Internet]. Madison Project Working Paper WP-15. [updated 2020; cited 2021 Jun 6]. Available from: https://www.rug.nl/ggdc/ historicaldevelopment/maddison/ publications/wp15.pdf

[14] Roser, M, Ortiz-Ospina, E. Global Extreme Poverty [Internet]. Published online at OurWorldInData.org. 2013. [cited 2021 Jun 6]. Available from: https://ourworldindata.org/ extreme-poverty

[15] World Bank. Poverty and shared prosperity 2018: Piecing together the poverty puzzle [Internet]. Washington, DC: World Bank. License: Creative Commons Attribution CC BY 3.0 IGO. [cited 2021 Jun 6]. Available from: https://openknowledge.worldbank.org/bitstream/handle/10986/30418/9781464813306.pdf

[16] IFPRI. Green Revolution: Curse or Blessing? [Internet]. International Food Policy Research Institute. Washington, D.C. 2002. [cited 2021 Jun 6]. https://oregonstate.edu/instruct/css/330/three/Green.pdf

[17] Saez, E. Striking it richer: The evolution of top incomes in the U.S [Internet]. Unpublished update of report published in Pathways Magazine, Stanford Center for the Study of Poverty and Inequality, Winter 2008, 6-7. U.C. Berkeley, Department of Economics. 2020. [cited 2021 Jun 6]. Available from: https://eml.berkeley.edu/~saez/saez-UStopincomes-2018.pdf

[18] Schoch, M, Lakner, C. The number of poor people continues to rise in Sub-Saharan Africa [Internet]. Published on Data Blog. World Bank. December 16, 2020. [cited 2021 Jun 6]. Available from: https://blogs.worldbank.org/opendata/number-poor-people-continues-rise-sub-saharan-africa-despite-slow-decline-poverty-rate

[19] Brosig, M. Frawley, P, Hill, A, Jahn, M, Marsicek, M, Paris, A, Rose, M, et al. Implications of climate change for the U.S. Army [Internet]. United States Army War College. Carlisle, PA; 2019. [cited 2021 Jun 6]. Available from: https://climateandsecurity.files.wordpress.com/2019/07/implications-of-climate-change-for-us-army_army-war-college_2019.pdf

[20] CASSE, 2021b. The downside of economic growth [Internet]. Center for the Advancement of the Steady State Economy. 2021. [cited 2021 Jun 6].

https://steadystate.org/discover/downsides-of-economic-growth/

[21] Brundtland, G. Report of the World Commission on Environment and Development: Our common future [Internet]. United Nations; 1987. United Nations General Assembly document A/42/427. [cited 2021 Jun 6]. Available from: https://sustainabledevelopment.un.org/content/documents/5987our-common-future.pdf

[22] United Nations Conference on Environment and Development (UNCED). Agenda 21, Rio Declaration, Forest Principles [Internet]. New York: United Nations; 1992. [cited 2021 Jun 6]. Available from: https://sustainabledevelopment.un.org/content/documents/Agenda21.pdf

[23] Hansen, Jay. Overshoot Loop: Evolution Under the Maximum Power Principle [Internet]. 2013. [cited 2021 Jun 6]. https://dieoff.com/page11.htm

[24] FRED. Federal Reserve Economic Data. Real Gross Domestic Product [Internet]. St. Louis Federal Reserve. 2021. [cited 2021 Jun 6]. https://fred.stlouisfed.org/series/GDPC1

[25] Stiglitz, J. GDP is the wrong tool for measuring what matters [Internet]. SciAm. 2020. August 1, 2020. [cited 2021 Jun 6]. Available from: https://www.scientificamerican.com/article/gdp-is-the-wrong-tool-for-measuring-what-matters/

[26] Tietenberg, T, Lewis, L. Environmental economics: the essentials. New York: Routledge. 2020.

[27] Pimentel, D; Wilson, C; McCullum, C; Huang, R; Dwen, P; Flack, J, et al. Economic and envi-ronmental benefits of biodiversity BioScience. 1997 Dec [cited 2021 Jun 29];(47)11:747-757. Available from: http://links.jstor.org/sici?sici=0006-3568%28199712%2947%3A11%3C747%3AEAEBOB%3E2.0.CO%3B2-H

[28] Quinney, M. 5 reasons why biodiversity matters – to human health, the economy and your wellbeing [Internet]. World Economic Forum; [cited 2021 Jun 29]. Available from: https://www.weforum.org/agenda/2020/05/5-reasons-why-biodiversity-matters-human-health-economies-business-wellbeing-coronavirus-covid19-animals-nature-ecosystems/

[29] Kyte, R. Crop diversity Is key to agricultural climate adaptation. Scientific American. Blog [Internet]. 2014 August 18, 2014. [cited 2021 Jun 29]. Available from: https://blogs.scientificamerican.com/guest-blog/crop-diversity-is-key-to-agricultural-climate-adaptation/

[30] Keesing, F., Belden, L., Daszak, P. et al. Impacts of biodiversity on the emergence and transmission of infectious diseases. Nature. 2010;(468):647-652. https://doi.org/10.1038/nature09575

[31] Wilkinson, D, Marshall, J, French, N, Hayman, D. Habitat fragmentation, biodiversity loss and the risk of novel infectious disease emergence. J R Soc Interface [Internet]. 2018 Dec 5 [cited 2021 Jun 29];15(149):20180403. Available from: https://pubmed.ncbi.nlm.nih.gov/30518565/ doi: 10.1098/rsif.2018.0403

[32] Neergheen-Bhujun, V, Taj Awan, A, Baran, Y, Bunnefeld, N, Chan, K, dela Cruz, T, et al. Bio-diversity, drug discovery, and the future of global health: Introducing the biodiversity to biomedi-cine consortium, a call to action. J Glob Health [Internet]. 2017 Dec [cited 2021 Jun 29];(7)2:020304. Available from: http://jogh.org/documents/issue201702/jogh-07-020304.pdf doi: http://jogh.org/documents/issue201702/jogh-07-020304.pdf

[33] Economic reasons for conserving wild nature. Balmford A, Bruner A, Cooper P, Costanza R, Farber S, Green R, et al. Science. 2002 Aug 09; (297)5583:950-953 DOI: 10.1126/science.1073947

[34] World Bank Open Data. Agriculture, forestry, and fishing, value added (constant 2010 US$) [Internet]. [cited 2021 Jun 29]. Available from: https://data.worldbank.org/indicator/NV.AGR.TOTL.KD

[35] BLS. 2020. Bureau of Labor Statistics. Net Domestic Product [Internet]. [cited 2021 Jun 6]. https://www.bea.gov/help/glossary/net-domestic-product-ndp

[36] Alfsen, KH, Hass, JL, Tao, H, You, W. International experiences with green GDP [Internet]. Statistics Norway. 2006. [cited 2021 Jun 6]. Available from: https://ise.unige.ch/isdd/IMG/pdf/Green_GDP_rapp_200632.pdf ISSN 0806-2056

[37] Talbarth J, Webb, J. Genuine progress indicator [Internet]. Green Growth Case Study Series. 2014. [cited 2021 Jun 6]. Available from: https://www.greengrowthknowledge.org/sites/default/files/downloads/best-practices/GGBP%20Case%20Study%20Series_United%20States_Genuine%20Progress%20Indicator.pdf

[38] Daly, H, Cobb, JB Jr. For the common good: redirecting the economy toward community, the environment, and a sustainable future. Boston: Beacon Press; 2012.

[39] Kubiszewski, I, Costanza, R, Franco, C, Lawn, P, Talberth, J, Jackson, T, Aylmer, C. Beyond GDP: Measuring and achieving global genuine progress [Internet]. EcolEcon. 93(5):57-68. [cited 2021 Jun 6]. Available from: https://doi.org/10.1016/j.ecolecon.2013.04.019 https://www-sciencedirect-com.ezproxy1.apus.edu/science/article/pii/S0921800913001584?via%3Dihub

[40] Almond, R, Grooten M., Petersen, T. (Eds). Living planet report 2020 - Bending the curve of biodiversity loss [Internet]. Gland, Switzerland, WWF. 2020. [cited 2021 Jun 29]. Available from: https://oursharedseas.com/oss_downloads/living-planet-report-2020-bending-the-curve-of-biodiversity-loss/

[41] OECD. OECD Better Life Index [Internet]. 2021. [cited 2021 Jun 6]. http://www.oecdbetterlifeindex.org/about/better-life-initiative/

[42] CASSE. Steady State Economy Definition [Internet]. Center for the Advancement of the Steady State Economy. 2021. [cited 2021 Jun 6]. https://steadystate.org/discover/definition/

[43] Kenton, Will. Steady-State Economy [Internet]. Investopedia. 2020. [cited 2021 Jun 6]. https://www.investopedia.com/terms/s/steady-state-economy.asp

[44] Raworth, K Doughnut economics : seven ways to think like a 21st-century economist [Internet]. London: Penguin Random House; 2017.

[45] Hardner, J, Rice R. Rethinking green consumerism. SciAm. 2002 May. 287:89-95.

[46] Niesten, E, Zurita, P, Banks, S. Conservation agreements as a tool to generate direct incentives for biodiversity conservation. Biodiversity. 2010. (11):5-8.

[47] CI. What on Earth is a 'Conservation Agreement' [Internet]. Conservation International. 2021. [cited 2021 Jun 6]. https://www.conservation.org/blog/what-on-earth-is-a-conservation-agreement

[48] CAF. Conservation Agreement Fund [Internet]. 2021. [cited 2021 Jun 6]. https://conservationagreementfund.org/projects/

[49] Rossetti, P, Bosch, D, Goldbeck, D. Comparing effectiveness of climate regulations and a carbon tax [Internet]. Unpublished research report. American Action Forum, Washington, D.C.; 2018. [cited 2021 Jun 6]. Available from: https://www.americanactionforum.org/research/comparing-effectiveness-climate-regulations-carbon-tax-123/#ixzz6wCB4GgcU

[50] Metcalf, GE. On the economics of a carbon tax for the United States [Internet]. Brookings Institution. Brookings Papers on Economic Activity. Spring 2019. [cited 2021 Jun 6]. Available from: https://www.brookings.edu/bpea-articles/on-the-economics-of-a-carbon-tax-for-the-united-states/

[51] Tessum, C., Hill, J. Marshall, D. Air quality impacts from light-duty transportation [Internet]. Proceedings of the National Academy of Sciences. 2014 Dec. 111 (52):18490-18495. [cited 2021 Jun 6]. Available from: https://www.pnas.org/content/111/52/18490 DOI: 10.1073/pnas.1406853111

[52] EPA. 2020. Sources of Greenhouse Gas Emissions [Internet]. [cited 2021 Jun 6]. https://www.epa.gov/ghgemissions/sources-greenhouse-gas-emissions#:~:text=Electricity%20production%20(25%20percent%20of,share%20of%20greenhouse%20gas%20emissions.

[53] UN-REDD Program. About REDD+ [Internet]. [cited 2021 Jun 6]. https://www.unredd.net/about/what-is-redd-plus.html

[54] Van der Werf, GR, Morton, DC, DeFries, RS, Olivier, CJ, Kasibhatla, PS, Jackson, RB, Collatz, CJ, Randerson, JT. CO2 emissions from forest loss [Internet]. NatGeosci. 2009. 2(11):737-738. [cited 2021 Jun 6]. Available from: https://escholarship.org/content/qt52n993mq/qt52n993mq.pdf

[55] CI. What on Earth is a 'REDD+'? [Internet]. Conservation International.

2021. [cited 2021 Jun 6]. https://www.
conservation.org/blog/what-on-
earth-is-redd

[56] Stern, NH. The economics of
climate change: the Stern review.
Cambridge, UK: Cambridge University
Press. 2007.

[57] Seymour, F, Busch, J. Why forests?
Why now? The science, economics, and
politics of tropical forests and climate
change [Internet]. Center for Global
Economic Development. 2016. ISBN:
978-1-933286-85-3.

[58] Roberts, D. Washington Votes No on
a Carbon Tax – Again [Internet]. Vox.
November 6, 2018. [cited 2021 Jun 6].
https://www.vox.com/energy-and-
environment/2018/9/28/17899804/
washington-1631-results-carbon-fee-
green-new-deal

[59] Seattle Times. Seattle Times
Recommends: No on Initiative 1631
[Internet]. 2018. [cited 2021 Jun 6].
https://www.seattletimes.com/opinion/
editorials/the-seattle-times-recommends-
no-on-initiative-1631/

[60] Meyers, M. Green bailouts: relying
on carbon offsetting will let polluting
airlines off the hook [Internet]. The
Conversation. 2020. [cited 2021 Jun 6].
https://theconversation.com/
green-bailouts-relying-on-carbon-
offsetting-will-let-polluting-airlines-
off-the-hook-137472

Chapter 3

Urban Ecosystem: An Interaction of Biological and Physical Components

Hassanali Mollashahi and Magdalena Szymura

Abstract

Urban ecosystems are composed of biological components (plants, animals, microorganisms, and other forms of life) and physical components (soil, water, air, climate, and topography) which interact together. In terms of "Urban Green Infrastructure (UGI)", these components are in a combination of natural and constructed materials of urban space that have an important role in metabolic processes, biodiversity, and ecosystem resiliency underlying valuable ecosystem services. The increase in the world's population in urban areas is a driving force to threat the environmental resources and public health in cities; thus, the necessity to adopt sustainable practices for communities is crucial for improving and maintaining urban environmental health. This chapter emphasizes the most important issues associated with the urban ecosystem, highlighting the recent findings as a guide for future UGI management, which can support city planners, public health officials, and architectural designers to quantify cities more responsive, safer places for people.

Keywords: urban green infrastructure, connectivity, ecosystem services, biodiversity, urban microbiome

1. Introduction

1.1 Urban ecosystem

Urban areas are composed of natural and constructed systems where the human population is more concentrated, and there are complex interactions between socioeconomic factors and biophysical processes [1, 2]. In a city, an ecological process often occurs in habitat patches, which are connected by corridors in a matrix of streets and buildings. The major ecological processes between/among habitat patches include immigration and dispersal agents, also, ecological corridors that can act as links or barriers for dispersal ability [2].

Due to transport networks cities are often the entry points of many alien species [3]. Moreover, in contrast with non-urban areas, urban ecosystems have different physical and chemical properties, which highly influence species distribution and ecosystems functioning [4, 5]. As a whole, urban areas have been usually considered novel in relation to their non-urban counterparts, which are comprised of a variety of fragmented habitats [4]. Overall, in this novel ecosystems the restoration ecology, conservation, biodiversity, ecosystem services, and climate change have been the most discussed topics in literature [6].

1.2 Urban green infrastructure

A bibliographic analysis of urban sustainability indicates that the topic of green infrastructure started to be in the attention of scientists in 2010, when, the awareness of issues associated with climate change was raised and the assessment of urban ecosystem services was more considered. During a period of five years (2010-2015), topics related to health and well-being were more interesting, and the motor theme of conversation became the priority of the scientists studying the importance of green infrastructures. This demonstrates the significant importance of green infrastructure and its association with sustainability [7, 8].

The term "Urban Green Infrastructure (UGI)" refers to engineered and non-engineered habitat structures in connection with natural and semi-natural areas and other environmental features, which are designed to deliver a wide range of services from nature to humans. Green infrastructure comprises different kinds of components (for example, parks, green roofs, urban forests, road verges) which according to several number of parameters (e.g., spatial scale, dimension, location) are categorized [9, 10].

The "Green Infrastructure" can perform several functions in the same spatial area. In contrast to gray (or conventional) infrastructure which usually has one single objective, GI is multifunctional which means it can promote win-win solutions or "small loss-big gain", delivering benefits to a wide range of stakeholders and the public at large [10].

In line with Europe's 2020 strategy, it can act as a catalyst for economic growth by inward investment and generating employment, reducing environmental costs, and providing health benefits among others. This can contribute to the recovery of Europe's economy by creating green businesses and innovative approaches, representing around 5% of the job market. For instance, the Hoge Kempen National Park (6,000 ha) which is located in the eastern part of Belgium, the investment to carry out improvement projects is raised up to €90 million and generating €24.5 million per year in revenues from sustainable tourism alone. In Sweden, 10,000 m^2 of green roofs were installed and an open storm-water system was built to improve the environment both for people and nature, the entire project cost around €22 million but the benefits that have been derived from this investment are already tracking up; for example, decreasing in rainwater runoff rates by half, significant saving energy by residents, increasing the biodiversity by half, unemployment has fallen from 30–6%, and turnover in tenancies is decreased substantially [10]. More example is Canada where the economic value of 13 ES in Canada's Capital Region (Ottawa-Gatineau region) amounts to an average of 332 million dollars, and to a total economic value of over 5 billion dollars, annualized over 20 years [11].

1.3 Ecosystem services

Improving the knowledge about the importance of urban ecosystem services (ESs), and their value especially in the current trend of world urbanization is necessary. Thus, the role of city planners and other disciplines and their collaboration to integrate new findings associated with ESs is necessary [12]. ESs, directly and indirectly, influence human life and thus the economic activities. For examples, the maintenance of soil fertility can secure food production, and/or providing clean air and water through the absorption of pollutants by plants, and our mental and physical health may depend on the accessibility to green spaces [13].

1.3.1 Categorization of ecosystem services (ES) at the urban level

We only consider the ecosystem services classified by the Mapping and Assessment of Ecosystems and their Services (or MAES), Urban ecosystem, 4th report (May 2016). This classification takes into account merely the ecosystem services which are more important and happen in urban areas. These ecosystem services (ESs) are including (i) provisioning services in which the food and water are the most valuable ones, (ii) regulating services including the regulation of air quality, flood and water flow regulation, also, noise and temperature reduction plus pollination, (iii) the cultural ecosystem services such as recreation, education and cultural heritage [14].

There is criticism this classification in which the supporting services is not taken into account. Those supporting services are so-called intermediate ecosystem services and comprise the habitats for species and maintenance of genetic diversity [15].

Apart from the above-mentioned classification system, the three other classifications are also available but they consider the assessment of ecosystem services on much big scale than cities. These three classifications are as follows; (1) CICES (the Common International Classification of Ecosystem Services), (2) The MA (the Millennium Ecosystem Assessment), and (3) TEEB (the Economics of Ecosystems and Biodiversity) [16–18].

1.4 Biodiversity

The urban area often contain threatened species. The spatial structure of the urban landscape, especially patches features (e.g., patch size and their connectivity) are correlated with species richness and biodiversity [19].

More than three-quarters of Earth species are characterized to be extinct at short time intervals which is unprecedented. Mammalia, birds, and amphibians are the groups of animals that have become more popular for the assessment by scientists [20], while insects species have been poorly studied, despite their vital role in ecosystems and in turn well-being. Biodiversity loss of insects is reported as a worldwide phenomenon, (typically in Great Britain and other European countries), where four main drivers of this condition have been presented [21, 22]. Habit lost and fragmentation which is made by the human is considered as the main factor of global biodiversity loss, and then pollution, biological factors, and climate change. In the case of mammals and birds, habit change plays the same role in the reduction of their species [23–25].

1.4.1 Biological factors

Human settlement and infrastructure development is a threat to protected species and negatively impact on the many of the at-risk species [26]. Among those species, beneficial insects like honeybee colonies, birds, and mammals are more endangered. For example, beehives are at risk of collapse by mite parasites and viral infection. Thus the necessity of conservation strategies is a need in urban wildlife, where the species encounter anthropized environments that differ from the natural landscape. With this in mind that many species characteristics such as dispersal ability, sex, even body mass influence the species movement to urban areas. Passerine birds are a good example; where the urban colonization rate of these birds is associated with the color dichromatism [27, 28].

If we consider two groups of specialized and generalist species, the first group (specialized) tend to be more susceptible and poor in adaptation to the habitat changes in novel conditions as they have a special host, and their ability to recover quickly is less; thus, these species are more at risk of extinction. The second group (generalized species) are more adaptable to climate change and can successfully colonize the new environment/urban setting in a short time, showing plasticity, adaptability, and having access to a wide range of food and shelter requirements. Other factors such as invasive species has been reported to show cascading effects on the ecosystem and influence the species communities, and the diversity of many organisms, especially insects. For example, cattle grazing and recreational activities negatively impacted the distribution of a dragonfly (*Ecchlorolestes peringueyi*) in South Africa [25].

1.4.2 Habitat change

Human activities like industrialization, and agricultural intensification, have changed the habitat structure of natural landscape, causing the reduction in food resources and shelter sites for many specialist species. Moreover, urbanization, causing the disappearance of many habitat specialists and their replacement with a few generalists adapted to the artificial human environment. Providing habitat quality and management contribute to biodiversity maintenance. A good example of habitat management is presented by Britain government where the area of flower grasslands was increased for the target populations of bumble species [25, 29].

1.4.3 Pollution

There are several factors causing environmental pollution, declining biodiversity loss. Fertilization and pesticide application mostly occur in agricultural settings. In the case of urban settings, industrial sites, transportation, and sewage increase soil contamination by the heavy metals in green infrastructures, which can reduce not only belowground biodiversity but also influences the vegetation structure of lawns and grasslands patches [25].

Several studies reported the existence of neonicotinoid residues that contaminated the honey samples from *Apis mellifera* hive collected from European honey samples. Neonicotinoids (e.g., Clothianidin, Imidacloprid, and Thiacloprid) have been identified from urban habitats, suggests the reconsideration of pesticide application in urban areas. Thus, due to urbanization and agricultural intensification the awareness should be raised about chronic toxicity and exposure of bees and other beneficial insects and consequently human health [30]. Fipronil is a pyrazole insecticide and is widely used in agricultural areas against larval Culex species. The toxicity of Fipronil has been found in urban runoff waters in California and showed acute toxicity to aquatic invertebrates in south-eastern Australia, suggested to cause disruption to aquatic ecosystems [31, 32]. The toxicity impacts of insecticides such as imidacloprid, bifenthrin, and fipronil are detected, causing the reduction in survival and feeding ability of black tiger shrimp (*Penaeus monodon*), which were also distributed in urban waterways [33]. Moreover, many other kinds of insecticides (e.g., Pyrethroid) have high toxicity on aquatic insects, crustaceans. Aquatic environments are more at risk of disruption where pesticide residues from agricultural and urban runoff are the major cause of biodiversity declines. Bifenthrin was found from urban runoff in river water, affecting the most important prey species for American River Chinook salmon which can cause a significant reduction in their abundance [34, 35].

In Germany, over the 27 years of study, about 80% of the flying insect biomass losses were caused by increases in pesticide application [36]. In a study in Paris, urbanization made a significant reduction in the population of the bird species called "House Sparrows" [37].

1.4.4 Climate change

Urban areas are under the pressures of population growth, urbanization and suburbanization processes, which interact with the climate, leading to the establishment of the urban climate. Urban climate is generally characterized by some particular features such as heat islands effects, dryness, urban flooding, cold, humidity and pollution, which can significantly affect human health [38]. Abiotic stress such as heat waves, drought, and flooding are the three most important factors, having not only socio-economic impacts but also constrains on global food security [39].

1.4.4.1 Urban climate, the heat-related phenomena, and its impact

The urban heat and its extreme impacts on social and environmental aspects on urban residents together with climatic change arising from global warming, alleviating agricultural crops, influencing the resiliency of the urban greenery and therefore a risk for human health. The heat-related phenomena are related to heatwaves and drought which produce negative effects as heat-related illness and heat-related mortality [40–42]. Triggering certain types of diseases have been reported due to hydro climatic treat and long-term exposure to heat-related stresses, for example; respiratory, gastrointestinal, caused by low humidity, high temperatures and lack of water for personal hygiene, and household cleaning [43].

1.4.4.2 Urbanization and sponge city concept

Water flooding is a serious problem in many cities of China. The concept of sponge city was developed for the first time in China in 2014 in order to deal with urban flooding and to attenuate urban runoff, and improve the purification in the concept of urban sustainability. The concept is being developed to make use of 'blue' and 'green' spaces in the urban environment to encourage stormwater management and control [44].

"The sponge city concept aims to (i) adopt and develop LID (low impact development) concepts which improve effective control of urban peak runoff, and to temporarily store, recycle and purify stormwater; (ii) to upgrade the traditional drainage systems using more flood-resilient infrastructure (e.g. construction of underground water storage tanks and tunnels) and to increase current drainage protection standards using LID systems to offset peak discharges and reduce excess stormwater; and (iii) to integrate natural water-bodies (such as wetlands and lakes) and encourage multi-functional objectives within drainage design (such as enhancing ecosystem services) whilst providing additional artificial water bodies and green spaces to provide higher amenity value". The integrating of mentioned targets with the management approaches envisaged to gradually solve urban water issues, providing esthetic services and other benefits for urban populations, and that to improve urban habitat based on nature-based solutions to maintain the biodiversity in cities environment. The sponge city concept has a lot of influence on the approaching socio-ecological issues, bringing together the ideas from different disciplines to tackle many challenges linked to water-related issues across the world [45].

1.4.4.3 Global warming and insect's decline

Global warming stimulates the decline of many beneficial insects, for example, wild bees and butterflies. However, global warming shows contrasting trends on the population density of butterflies in Finland. Despite this, the general trend of the world's insect population exhibiting around 50 percent reduction. Likewise, the insect populations which are adapted to the cold climate have declined (e.g. dragon-flies, stoneflies, and bumblebees), showing a general reduction in population density of pollinators such as wasps, ants, and beetles in Mediterranean regions [25, 46].

1.5 Connectivity

Connectivity is demonstrated to be a proxy for biodiversity, where species and other ecological flows are able to move through a landscape and gain diversity in their genetic structure, stabilizing the ecosystem. As a result of urbanization, habitat fragmentation leading to the extinction of the threatened species, making the network between urban green infrastructure more important. Therefore, mod-eling the connectivity between different urban patches in an urban area through designing green corridors is stated to be a realistic direction. Connectivity has two elements; structural and functional connectivity in which the structural connectiv-ity is a useful indicator of functional connectivity, providing information on how to create a better connectedness of urban green spaces [47]. Different methods have been used to analyze the connectivity in an urban landscape. The graph theory method is the most useful tool by which the two concepts of inter and intra-patch connectivity is taken into account. This method is a robust metric, enabling to prioritization of the importance of each patch in the entire system [48].

1.5.1 Connectivity indices

Connectivity has three indices; (i) Number of links (L) between/among habitat patches (node) which provide information about the geographical distance between/ among patches, showing the physical structure between patches, (ii) number of components (NC), where a component is a set of patches/nodes which are connected by links; a patch itself is also considered as a component, and (iii) the integral index of connectivity (IIC), which was proposed by Pascual-Hortal and Saura [48, 49]. The connectivity raises when the NL is higher and the NC is lower. Considering IIC, the degree of connectivity within a landscape can be estimated, and also the contribu-tion of each patch into entire landscape connectivity which is the most useful tool, providing significant conceptual improvements in the decision process for planning [50–52]. The IIC shows the importance of every single patch in the overall connectiv-ity which is based on graph structure and binary connection model, which means two patches are connected or not. Assessing this index is based on delta/d (dIIC) or the differences in the IIC value and ranges from 0 to 1 for each patch, indicating the importance of each patch with a higher value in the overall connectivity of the analyzed landscape. The dIIC value has three fractions and each fraction additively leads to the overall value. The three fractions are including $dIIC_{intra}$ or intra-patch connectivity, $dIIC_{flux}$ or inter-patch connectivity when a patch is directly connected to the other one; $dIIC_{connector}$ or stepping stone, which means if a patch/node contrib-ute to the connection of other patches [53].

1.6 Urban microbiome

Microorganisms are a vital component of nature and can be found everywhere or so-called ubiquitous, from the human gut to natural ecosystems like oceans. They belong to bacteria, fungi, viruses, and micro-eukaryotes [54, 55]. In terms of envi-ronment, soil microbial communities are a key factor in the biochemical processes that support plant growth and other ecosystem services of GI features [56, 57]. At the urban level, the first assessment of subsurface microbial communities in a truly urban site was investigated in 1992 [58].

Edaphic variables are the factors related to the soil properties (e.g., soil pH) that affect the diversity and geographical distribution of microorganisms like soil bacterial communities; soil with lower pH (>4.5) has lower bacterial diversity [59]. As, in urban areas, the soil physical (moisture and texture) and chemical properties (pH, solid minerals, and organic matter) can influence microorganism communities [60, 61]. Notably, bacterial diversity is significantly correlated with human population density (as a proxy of anthropogenic activity) [62], indicating co-occurrence of human settlements and species-rich regions [63]; the reason for this relationship is unknown.

The results of human activities including heavy metals and other pollutants such as pesticides, fertilizers, salt, exposure to petroleum products impact the soil ecosystem, as these activities and products can alter the structure of soil bacteria communities and have a strong effect on their abundance and diversity [64–66].

Different urban soil types and their locations show that the Phyla Acidobacteria and Actinobacteria, are the most dominant soil bacteria [67]. On the other side, the most abundant fungi are related to the genera Glomus and Rhizophagus. The identified taxa are able to survive in distributed habitats and are associated with key ecosystem services (for example, decomposition and N cycling) [68].

Knowing microbial communities in GI features is important because it can help to guide urban planning for the purposes of improving urban biodiversity or bioremediation as a guide for future GI management. Identifying and understanding the dynamics of microbial communities in urban environments is thus essential for managing microbes beneficially in the context of urban sustainability [69]. Recently and in 2016 the project of Metagenomics and Meta-design of the Subways and Urban Biomes (MetaSUB) have started to characterize the composition of the microbial inhabitants of urban environments across the world. The aim of this international project is to support city planners, public health officials, and architectural designers and to quantify cities more responsive, safer places for people [70].

Growing the world's population accelerates the increase of pollutants and consequently can jeopardize the people's life by being exposure to pollutants. This can also proliferate the spread of pandemic and pathogenic microbiome. Therefore, it is imperative to adopt sustainable practices and enhance the health of the urban environment, considering the implementation of surveillance programs, discovering the genetic characterization and functional diversity of microbes in the cities [71, 72].

2. Conclusion

This chapter attempts to address the important concepts related to urban ecosystem. Urban areas are composed of natural and constructed systems. In a city, an ecological process including immigration and dispersal agents often occur in habitat patches, which are connected by corridors. Urban ecosystems have different physical and chemical properties, which highly influence species distribution, ecosystems functioning, and provide ample ecosystem services, representing sustainable tourism, saving energy, increasing the biodiversity, reducing environmental costs and providing health benefits for residents. Nowadays, however, urban development threatens human health and some elements of biodiversity, which is mainly caused by climate change especially urban heat island, environmental pollution, and habitat fragmentation. Green corridor is proposed to be pragmatic approach in connectedness of different groups of habitat structures and in turn genetic diversity. Subsurface microbial communities are also associated with major biochemical process which support plant growth and ensure key ecosystem services involving nitrogen cycling, biodegradation, and decomposition.

In an increasing urbanized world, adopting sustainable practices for communities are crucial for improving and maintaining urban environmental health. This could be helpful to guide urban planning for the purposes of improving urban biodiversity or bioremediation as a guide for future GI management. To do this, researchers from different disciplines, both in national and international collaborations can address many environmental issues and consequently human well-being in cities. To explore next, multidisciplinary, interdisciplinary, transdisciplinary projects are required to untangle the current challenges associated with biodiversity, ecosystem services, and climate change in urban areas.

Acknowledgements

This chapter financed by the project "UPWR 2.0: international and interdisciplinary programme of development of Wrocław University of Environmental and Life Sciences", co-financed by the European Social Fund under the Operational Program Knowledge Education Development, under contract No. POWR.03.05.00-00-Z062/18.

Author details

Hassanali Mollashahi* and Magdalena Szymura
Institute of Agroecology and Plant Production, Wrocław University of Environmental and Life Sciences, Wrocław, Poland

*Address all correspondence to: hassanali.mollashahi@upwr.edu.pl; hassanali.mollasahi@gmail.com

IntechOpen

References

[1] MEA (2005) Urban systems. Ecosystems and human well-being: current state and trends. Island Press, Washington, DC, pp 795-825

[2] Schaefer VH (2011) Remembering our roots: a possible connection between loss of ecological memory, alien invasions and ecological restoration. Urban Ecosyst 14:35-44.

[3] Gaertner M, Wilson JRU, Cadotte MW et al (2017) Non-native species in urban environments: patterns, processes, impacts and challenges. Biol Invasions 19:3461-3469.

[4] Kowarik I (2011) Novel urban ecosystems, biodiversity, and conservation. Environ Pollut 159:1974-1983.

[5] Perring MP, Manning P, Hobbs RJ et al (2013a) Novel urban ecosystems and ecosystem services. In: Hobbs RJ, Higgs ES, Hall CM (eds) Novel ecosystems: intervening in the new ecological world order. Wiley, Chichester, pp. 310-325.

[6] Teixeira, C.P. and Fernandes, C.O., 2020. Novel ecosystems: a review of the concept in non-urban and urban contexts. Landscape Ecology, 35(1), pp.23-39.

[7] Sharifi, A., 2020. Urban sustainability assessment: An overview and bibliometric analysis. *Ecological Indicators*, p.107102.

[8] Wang, J., Banzhaf, E., 2018. Towards a better understanding of Green Infrastructure: a critical review. Ecol. Ind. 85, 758-772.

[9] Deeb, M., Groffman, P. M., Joyner, J. L., Lozefski, G., Paltseva, A., Lin, B., et al. (2018). Soil and microbial properties of green infrastructure stormwater management systems.

Ecological Engineering 125, 68-75. doi: 10.1016/j.ecoleng. 2018.10.017

[10] European Commission. (2013). Building a Green Infrastructure for Europe. https://doi. org/10.2779/54125.

[11] L'Ecuyer-Sauvageau, C., Dupras, J., He, J., Auclair, J., Kermagoret, C. and Poder, T.G., 2021. The economic value of Canada's National Capital Green Network. *Plos one*, 16(1), p.e0245045.

[12] Russo, A., & Cirella, G. T. (2021). Urban Ecosystem Services: New Findings for Landscape Architects, Urban Planners, and Policymakers.

[13] Andreucci, M.B.; Russo, A.; Olszewska-Guizzo, A. Designing Urban Green Blue Infrastructure for Mental Health and Elderly Wellbeing. Sustainability 2019, 11, 6425. [CrossRef]

[14] European Commission. Directorate General for the Environment. In Mapping and Assessment of Ecosystems and Their Services; Urban Ecosystems, 4th Report; Publications Office: Luxembourg, 2016.

[15] Croci, E., Lucchitta, B. and Penati, T., 2021. Valuing Ecosystem Services at the Urban Level: A Critical Review. Sustainability 2021, 13, 1129.

[16] Haines-Young, R.; Potschin, M. CICES Version 4: Response to Consultation; Centre for Environmental Management, School of Geography, University of Nottingham: Nottingham, UK, 2012; p. 17.

[17] Reid,W.; Mooney, H.; Cropper, A.; Capistrano, D.; Carpenter, S.; Chopra, K. Millennium Ecosystem Assessment. Ecosystems and Human Well-Being: Synthesis; Island Press: Washington, DC, USA, 2005.

[18] Sukhdev, P.;Wittmer, H.; Schröter-Schlaack, C.; Neßhöver, C.; Bishop, J.;

Ten Brink, P.; Gundimeda, H.; Kumar, P.; Simmons, B. Mainstreaming the Economics of Nature: A Synthesis of the Approach, Conclusions and Recommendations of TEEB; UNEP: Nairobi, Kenya, 2010.

[19] Grafius, D. R., Corstanje, R., Warren, P. H., Evans, K. L., Norton, B. A., Siriwardena, G. M.,... & Harris, J. A. (2019). Using GIS-linked Bayesian Belief Networks as a tool for modelling urban biodiversity. Landscape and Urban Planning, *189*, 382-395.

[20] Barnosky, A.D., Matzke, N., Tomiya, S., Wogan, G.O.U., Swartz, B., Quental, T.B., Marshall, C., McGuire, J.L., Lindsey, E.L., Maguire, K.C., Mersey, B., Ferrer, E.A., 2011. Has the Earth's sixth mass extinction already arrived? Nature 471, 51-57.

[21] Fox, R., 2013. The decline of moths in Great Britain: a review of possible causes. Insect Conserv. Divers. 6, 5-19.

[22] Thomas, J.A., Telfer, M.G., Roy, D.B., Preston, C.D., Greenwood, J.J.D., Asher, J., Fox, R., Clarke, R.T., Lawton, J.H., 2004. Comparative losses of British butterflies, birds, and plants and the global extinction crisis. Science 303, 1879-1881.

[23] Chamberlain, D.E., Fuller, R.J., 2000. Local extinctions and changes in species richness of lowland farmland birds in England and Wales in relation to recent changes in agricultural land-use. Agric. Ecosyst. Environ. 78, 1-17.

[24] Diamond, J.M., 1989. The present, past and future of human-caused extinctions. Philos. Trans. R. Soc. Lond. Ser. B Biol. Sci. 325, 469-477.

[25] Sánchez-Bayo, F. and Wyckhuys, K.A., 2019. Worldwide decline of the entomofauna: A review of its drivers. Biological conservation, *232*, pp.8-27.

[26] McDonald RI, Kareiva P, Forman RTT (2008) The implications of current and future urbanization for global protected areas and biodiversity conservation. Biological Conservation, 141, 1695-1703.

[27] Evans, K. L., Chamberlain, D. E., Hatchwell, B. J., Gregory, R. D., & Gaston, K. J. (2011). What makes an urban bird?. Global Change Biology, *17*(1), 32-44.

[28] Iglesias-Carrasco, M., Duchêne, D. A., Head, M. L., Møller, A. P., & Cain, K. (2019). Sex in the city: sexual selection and urban colonization in passerines. Biology letters, *15*(9), 20190257.

[29] Williams, P., Osborne, J., 2009. Bumblebee vulnerability and conservation world-wide. Apidologie 40, 367-387.

[30] Kavanagh, S., Henry, M., Stout, J. C., & White, B. (2021). Neonicotinoid residues in honey from urban and rural environments. Environmental Science and Pollution Research, 1-12.

[31] Stevens, M., Burdett, A.S., Mudford, E., Helliwell, S., Doran, G., 2011. The acute toxicity of fipronil to two non-target invertebrates associated with mosquito breeding sites in Australia. Acta Tropica 117, 125-130.

[32] Gan, J., Bondarenko, S., Oki, L., Haver, D., Li, J., 2012. Occurrence of fipronil and its biologically active derivatives in urban residential runoff. Environ. Sci. Technol. 46, 1489-1495.

[33] Hook, S.E., Doan, H., Gonzago, D., Musson, D., Du, J., Kookana, R., Sellars, M.J., Kumar, A., 2018. The impacts of modern-use pesticides on shrimp aquaculture: An assessment for north eastern Australia. Ecotoxicol. Environ. Saf. 148, 770-780.

[34] Beketov, M.A., Liess, M., 2008. Acute and delayed effects of the neonicotinoid insecticide thiacloprid on

seven freshwater arthropods. Environ. Toxicol. Chem. 27, 461-470.

[35] Weston, D.P., Schlenk, D., Riar, N., Lydy, M.J., Brooks, M.L., 2015. Effects of pyrethroid insecticides in urban runoff on Chinook salmon, steelhead trout, and their invertebrate prey. Environ. Toxicol. Chem. 34, 649-657.

[36] Hallmann, C.A., Sorg, M., Jongejans, E., Siepel, H., Hofland, N., Schwan, H., Stenmans, W., Müller, A., Sumser, H., Hörren, T., Goulson, D., de Kroon, H., 2017. More than 75 percent decline over 27 years in total flying insect biomass in protected areas. PLoS One 12, e0185809.

[37] Mohring, B., Henry, P. Y., Jiguet, F., Malher, F., & Angelier, F. (2020). Investigating temporal and spatial correlates of the sharp decline of an urban exploiter bird in a large European city. Urban Ecosystems, 1-13.

[38] Grigorescu, I., Mocanu, I., Mitrică, B., Dumitrașcu, M., Dumitrică, C., & Dragotă, C. S. (2021). Socio-economic and environmental vulnerability to heat-related phenomena in Bucharest metropolitan area. Environmental Research, *192*, 110268.

[39] Fahad, S., Bajwa, A.A., Nazir, U., Anjum, S.A., Farooq, A., Zohaib, A., Sadia, S., Nasim, W., Adkins, S., Saud, S., Ihsan, M.Z., Alharby, H., Wu, C., Wang, D., Huang, J., 2017. Crop production under drought and heat stress: plant responses and management options. Front. Plant Sci. 8, 1147. https://doi.org/10.3389/fpls.2017.01147.

[40] Leal Filho, W., Icaza, L.E., Neht, A., Klavins, M., Morgan, E.A., 2018. Coping with the impacts of urban heat islands. A literature based study on understanding urban heat vulnerability and the need for resilience in cities in a global climate change context. J. Clean. Prod. 171, 1140-1149.

[41] Yoo, S., 2019. Assessing urban vulnerability to extreme heat-related weather events. The Routledge Handbook of Urban Resilience. In: Burayidi, M.A., Allen, A., Twigg, J., Wamsle, C. (Eds.). Routledge International Handbooks, p. 534.

[42] Graczyk, D., Kundzewicz, Z.W., Chory'nski, A., Førland, E.J., Pi'nskwar, I., Szwed, M., 2019. Heat-related mortality during hot summers in Polish cities. Theor. Appl. Climatol. 136 (3-4), 1259-1273.

[43] Ortega-Gaucin, D., de la Cruz Bartoľon, J., Castellano Bahena, H., 2018. Drought vulnerability indices in Mexico. Water 10 (11), 1671.

[44] MINISTRY OF HOUSING AND URBAN-RURAL DEVELOPMENT 2014. The construction guideline of sponge city in China – low impact development of stormwater system (trail) In: DEVELOPMENT, M. O. H. A. U.-R. (ed.). Beijing.

[45] Chan, F. K. S., Griffiths, J. A., Higgitt, D., Xu, S., Zhu, F., Tang, Y. T.,... & Thorne, C. R. (2018). "Sponge City" in China—A breakthrough of planning and flood risk management in the urban context. Land use policy, *76*, 772-778.

[46] Stefanescu, C., Aguado, L.O., Asís, J.D., Baños-Picón, L., Cerdá, X., García, M.A.M., Micó, E., Ricarte, A., Tormos, J., 2018. Diversidad de insectos polinizadores en la península ibérica. Ecosistemas 27, 9-22.

[47] Zhang, Z., Meerow, S., Newell, J. P., & Lindquist, M. (2019). Enhancing landscape connectivity through multifunctional green infrastructure corridor modeling and design. Urban forestry & urban greening, 38, 305-317.

[48] Pascual-Hortal, L., & Saura, S. (2006). Comparison and development of new graph-based landscape connectivity indices: towards the

priorization of habitat patches and corridors for conservation. Landscape ecology, *21*(7), 959-967.

[49] Rayfield B, Fortin MJ, Fall A. Connectivity for conservation: a framework to classify network measures. Ecology. 2011 Apr; 92(4):847-58. https://doi.org/10.1890/09-2190.1 PMID: 21661548

[50] Urban D, Keitt T. Landscape connectivity: a graph-theoretic perspective. Ecology. 2001 May; 82(5): 1205-1218.

[51] Jalkanen J, Toivonen T, Moilanen A. Identification of ecological networks for land-use planning with spatial conservation prioritization. Landscape Ecology. 2020 Feb; 35(2):353-371.

[52] Matos C, Petrovan SO, Wheeler PM, Ward AI. Landscape connectivity and spatial prioritization in an urbanising world: A network analysis approach for a threatened amphibian. Biological Conservation. 2019 Sep 1; 237:238-247.

[53] Saura S, Rubio L. A common currency for the different ways in which patches and links can contribute to habitat availability and connectivity in the landscape. Ecography. 2010 Jun; 33 (3): 523-537.

[54] Human Microbiome Project Consortium. Structure, function and diversity of the healthy human microbiome. Nature. 2012;486:207-214.

[55] Sunagawa S, Coelho LP, Chaffron S, Kultima JR, Labadie K, Salazar G, et al. Structure and function of the global ocean microbiome. Science. American association for the. Adv Sci. 2015;348:1261359

[56] Wall, D. H. (2004). Sustaining Biodiversity and Ecosystem Services in Soils and Sediments. Washington, DC: Island Press.

[57] Hostetler, M., Allen, W., and Meurk, C. (2011). Conserving urban biodiversity? Creating green infrastructure is only the first step. Landscape and Urban Planning 100, 369-371. doi: 10.1016/j. landurbplan.2011.01.011

[58] Madsen, E.L., Winding, A., Malachowsky, K., Thomas, C.T. and Ghiorse, W.C., 1992. Contrasts between subsurface microbial communities and their metabolic adaptation to polycyclic aromatic hydrocarbons at a forested and an urban coal-tar disposal site. Microbial ecology, *24*(2), pp.199-213.

[59] Fierer, N.; Jackson, R. B. The diversity and biogeography of soil bacterial communities.Proc. Natl. Acad. Sci. USA 2006, *103* (3), 626-631.

[60] Brodsky, O.L., Shek, K.L., Dinwiddie, D., Bruner, S.G., Gill, A.S., Hoch, J.M., Palmer, M.I. and McGuire, K.L., 2019. Microbial communities in bioswale soils and their relationships to soil properties, plant species, and plant physiology. Frontiers in microbiology, *10*, p.2368.

[61] Joyner, J. L., Kerwin, J., Deeb, M., Lozefski, G., Prithiviraj, B., Paltseva, A., et al. (2019). Green infrastructure design influences communities of urban soil bacteria. Front. Microbiol. 10:14. doi: 10.3389/fmicb.2019.00982

[62] Wang, H., Cheng, M., Dsouza, M., Weisenhorn, P., Zheng, T. and Gilbert, J.A., 2018. Soil bacterial diversity is associated with human population density in urban greenspaces. Environmental science & technology, *52*(9), pp.5115-5124.

[63] Luck, G. W. A review of the relationships between human population density and biodiversity. Biol Rev Camb Philos Soc 2007, *82* (4), 607-645.

[64] Marcin, C., Marcin, G., Justyna, M.-P., Katarzyna, K., and Maria, N.

(2013). Diversity of microorganisms from forest soils differently polluted with heavy metals. Applied Soil Ecology 64, 7-14. doi: 10.1016/j.apsoil.2012. 11.004

[65] Delgado-Balbuena, L., Bello-López, J. M., Navarro-Noya, Y. E., Rodríguez-Valentín, A., Luna-Guido, M. L., and Dendooven, L. (2016). Changes in the Bacterial Community Structure of Remediated Anthracene-Contaminated Soils. PLoS ONE 11:e160991–e160928. doi: 10.1371/journal.pone.0160991

[66] Adeniji, A. O., Okoh, O. O., and Okoh, A. I. (2017). Petroleum Hydrocarbon Fingerprints of Water and Sediment Samples of Buffalo River Estuary in the Eastern Cape Province, South Africa. J Anal Methods Chem 2017, 2629365-2629313. doi: 10.1155/2017/2629365

[67] Huot, H., Joyner, J., Córdoba, A., Shaw, R. K., Wilson, M. A., Walker, R., et al. (2017). Characterizing urban soils in New York City: profile properties and bacterial communities. J Soils Sediments 17, 393-407. doi: 10.1007/s11368-016-1552-9

[68] McGuire, K. L., Payne, S. G., Palmer, M. I., Gillikin, C. M., Keefe, D., Kim, S. J., et al. (2013). Digging the New York City Skyline: Soil Fungal Communities in Green Roofs and City Parks. PLoS ONE 8:e58020–e58013. doi: 10.1371/journal.pone. 0058020

[69] King, G. M. (2014). Urban microbiomes and urban ecology: how do microbes in the built environment affect human sustainability in cities?. Journal of Microbiology, 52(9), 721-728.

[70] MetaSUB International Consortium. The metagenomics and Metadesign of the subways and urban biomes (MetaSUB) international consortium inaugural meeting report. Microbiome. 2016;4:24.

[71] Gardy JL, Loman NJ. Towards a genomics-informed, real-time, global pathogen surveillance system. Nat Rev Genet. 2018;19:9-20.

[72] Miller RR, Montoya V, Gardy JL, Patrick DM, Tang P. Metagenomics for pathogen detection in public health. Genome Med. 2013;5:81.

Chapter 4

The Diversity of Endophytic *Aspergillus*

Vo Thi Ngoc My and Nguyen Van Thanh

Abstract

In plants, endophytic fungi and plant are closely related, endophytic fungi can use substances in plants as nutrients to survive. In return, they bring many benefits to the plant, playing an important role in protecting the host plant against the harmful effects of insects, harmful microorganisms or environmental disadvantages. Recently, secondary fungi metabolites, especially endophytic fungi, are gaining interest because they can produce many bioactive metabolites with antibacterial, anticancer and antioxidant properties. Some endophytic fungi are noted as *Aspergillus, Penicllium, Fusarium* due to the production of many metabolites for biological effects such as antibacterial, antiviral, anticancer, etc. in which *Aspergillus* species product many compounds have properties antibacterial such as terrequinon A, terrefuranon, Na-acetyl aszonalemin, etc.

Keywords: endophytic fungi, *Aspergillus*, endophyte, secondary metabolites, biological activity

1. Introduction

1.1 Endophytic microorganisms

Endophytic microorganisms are microorganisms that live in the plant tissue beneath the epidermal cell layers without harming or infecting the host plant, endophytic microorganisms that live in the intercellular space of tissues and thereby they can invade living cells [1].

1.2 Endophytic fungi

Endophytic fungi account for a high percentage of the current group of endophytic plant microorganisms. They are considered a source of many new substances, including many active substances with interesting biological effects. These fungal forms can be detected incidentally in the deep tissues of normally growing host plants. They are endogenous to the host plant and, thanks to their strong biosynthetic capacity, are able to produce a large number of metabolites. This may lead to the emergence of new bioactive substances and promises to develop production on an industrial scale. In addition, the substances produced by endophytic fungi are considered as an agent to help balance the microflora on the host plant to prevent pathogens [2, 3].

1.3 Relationship between endophytic fungi and plants

Endophytic fungi can be easily isolated from a surface-sterilized piece of plant tissue. The number of endophytic fungi found was also very variable when examining different plant samples, this number can range from one to several hundred strains.

The presence of endophytic fungi in plant tissues can be explained in many different ways. But perhaps the most plausible is the hypothesis that endophytic fungi arose from some plant pathology in the evolution of plants. The tree also has a microflora, in which there are strains that exist dormant and cause disease only when the tree is old and weak or facing adverse living conditions. The interaction between the host plant and the pathogenic microorganism during long-term development has resulted in genetic mutations from the pathogenic microorganisms to yield useful strains of endophytic fungi without causing disease [1].

Between endophytic fungi and host plants there is a symbiotic relationship, mutualism or mutualism, etc. The symbiotic or mutualistic relationship between endophytic fungi and plants is shown quite closely. At times they are closely linked as a single individual and contribute to the distinctive character of the tree.

Endophytic fungi promote ecological adaptation of host plants. In some plant species with endophytic fungi, it has been found that they have increased drought tolerance or tolerance of aluminum toxicity in water sources, in habitats, etc. In addition to protecting plants against a number of factors detrimental to the host such as herbivores or insects, many natural products produced by endophytic fungi have also been observed, monitored and concluded on the ability to prevent, inhibit or kill many different pathogens that invade plant tissues. That is also the reason why some endogenous fungal strains can produce phytochemicals that give the host plant a unique and distinctive character [3, 4].

For example: in the early 1990s, a novel taxol-producing endophytic fungus, *Taxomyces andreanae*, was isolated from *Taxus brevifolia*. This set the stage for a more comprehensive examination of the ability of other *Taxus* species and other plants to yield endophytes producing taxol. There is an endophytic fungi strain that produces taxol, an important active substance of great significance in the field of medicine.

1.4 Basic principles for selecting plants to isolate endophytic fungi

Young plant tissues are more suitable for isolating endophytic fungi than mature tissues because adult tissues often contain many different types of fungi, making isolation difficult. Collected plant samples should be stored at 4°C until endophytic fungi are isolated, and isolation should be carried out as soon as possible to avoid airborne bacterial contamination.

To obtain endophytic fungi with biological activity, it is necessary to select plant species that are outstanding in terms of biology, age, endemism, botanical history, and habitat of the host plant. Many studies have shown that medicinal plants and plants living in special environments are frequently studied to screen for endophytic fungi that produce antibiotic substances [1].

1.4.1 Plants live in unusual biological environments

With these plants, the unusual environment and harsh natural conditions force the tree to survive, a special element is needed to make the plant highly resistant. And one would expect that factor to be beneficial endophytic fungi [4].

For example: showed that an aquatic plant, *Rhyncholacis penicillata*, which lives in harsh aquatic environment which may be constantly wounded by passing rocks and other debris, resists infection by common oomyceteous fungi that cause disease. The possibility that endophytes associated with this aquatic plant may produce antifungal agents that protect the plant from attack by pathogenic fungi is feasible. A novel antioomycetous compound, oocydin A was discovered from the endophytic strain *Serratia marcescens* from this plant.

1.4.2 Plants are used as folk medicine

A number of plants have been used according to folk experience, from generation to generation for wound healing, antifungal, antibacterial, etc.

For example: A study of endophytic fungi producing novel bioactive substances in Brazil. It is a plant named "Mexican Sunflower" - *Tithonia diversifolia* (Asteraceae) that is often mentioned with many interesting points. For a long time, based on oral experience, people have used this plant to cure a number of diseases such as malaria, diarrhea, viral fever, hepatitis and to heal open wounds. In addition, extracts from *T.diversifolia* with anti-inflammatory, amoebic, antispasmodic, antifungal, antibacterial, and antiviral activities were also mentioned. Based on that information, scientists have isolated and isolated *Phoma sorghina*, an endogenous fungus from the leaves of this "Mexican sunflower". From this, six anthraquinone derivatives were obtained from fungal metabolites, half of which are new known substances [1].

1.4.3 Ecologically specific plants

Plants with unusually long lifespans, growing in areas of great biological change, or living in ancient soils are also ideal research subjects to provide endogenous fungi new. Plants surrounded by plants infected with the pathogen, but not infected, are more likely to harbor endophytic fungi with antimicrobial activity than other host plants.

For example: 2008, Tuntiwachwuttikul et al., found an endogenous fungal that was active against pathogens on banana plant *Colletotrichum musae* (Phyllachoraceae) [5].

1.4.4 Endemic plants

Endemic plant species that have a normal lifespan, or occupy a certain area of land in the wild. *Chaetomium globosum* isolated from leaves of endemic plant *Maytenus hookeri* distributed only in Yunnan regions, China produces chaetoglobosin B which inhibits tuberculosis bacteria [5].

1.5 Diversity of endophytic fungi

Endophytic fungi are very abundant, according to a study by Matsushima in 1971 conducted on some angiosperms and conifers in North America and Panama, which suggested that most of the endophytic fungi belong to the *Ascomycetes* class [4].

2008, Huang et al. also found endogenous fungi present in 27/29 surveyed medicinal plants. The frequency of occurrence of endophytic fungi is relatively high, mainly the genera *Fusarium* (27%), *Colletotrichum* (20%), *Phomopsis* (11%), *Alternaria* (9%), *Aspergillus* (5%), etc. **Figure 1**.

According to the statistics of scientists studying three plant families in Southeast Arizona - USA, forests in North Carolina and Northern forests shows that:

- Surveying the host plant representing the Fagaceae family obtained 44 endogenous fungal strains in which the genus *Sordariomycetes* predominated.

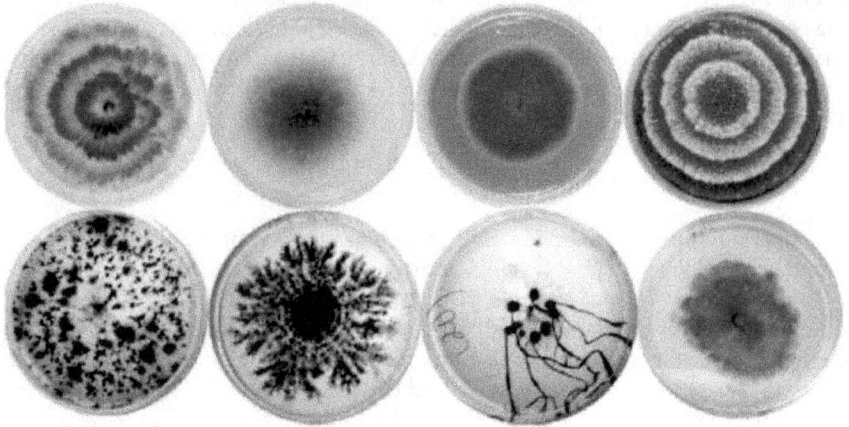

Figure 1.
Strains of class Ascomycetes isolated from leaves of angiosperms and conifers in North America and Panama (source: Selim KA et al., 2012).

- Surveying *Pinus ponderosa* trees representing the Pinaceae family obtained 111 endogenous fungal strains, in which the genus *Leotiomycetes* predominated.

- Surveying plants *Cupressus arizonica* and *Platycladus orientalis* representing the Cupressaceae family obtained 42 strains, in which the genus *Dothideomycetes* prevailed [4].

1.6 Endogenous fungi produce biologically active substances

Many endogenous fungi in plants have been isolated, they have the ability to produce biologically active substances such as antibiotic, antibacterial, antifungal, tumor suppressor, antioxidant and other biological activities.

1.6.1 Endogenous fungi with antibacterial and antifungal properties

Many studies on antibacterial and antifungal activities are produced from endogenous fungi, mainly belonging to the following groups: alkaloids, peptides, steroids, terpenoids, phenols, quinines and flavonoids, etc. These compounds account for only a part small in the total number of active substances produced by endophytic fungal species, they are clearly an excellent and novel potential source for the production of new antibiotics. This holds great promise for solving the problem of drug resistance in bacteria because these antibiotics are novel and highly active compounds **Table 1**.

Besides, *Phomopsis* is an endogenous fungus that produces phomopsichalasin, representing the first group of compounds of cytochalasin type.

1.6.2 Endogenous fungal tumor suppressor

Since 1990, *Taxomyces* andreanae was first isolated from *T. brevifolia*, this fungus produces paclitaxel - an inhibitor of achromatic spindles during cell division, with a mass spectrum similar to paclitaxel extracted from *Taxus*. Subsequently, several paclitaxel-producing fungi have been isolated from many plants and other *Taxus* species **Table 2**.

Endophytic fungi	Isolation source	Antibiotic	Antimicrobial effects	Ref.
Pestralotiopisis microspora	*Torreya axifolia*	Pestalopyrone Hydroxypestalopyrone	Plant antibiotics	[4, 6]
Pestralotiopisis jesteri	Trees that grow in the rivers of Papua New Guinea	Hydroxyl jesterone	Antifungal plant disease	[7]
Colletotrichum gloeosporioides	*Artemisia mongolica*	Acid colletotric	Antimicrobial *Helminthosporium sativum*	[4]
Colletotrichum sp.	*Artemisia annua*	Metabolites	Antibacterial, anti-fungal that causes diseases for humans and plants	[4]
Muscodor albus	*Cinnamomum zeylanicum*	Volatile matter	Inhibits bacteria and fungi	[4]
Fusarium sp.	*Selaginella pallescens*	CR 337: New pentaketide	*Candida albicans*	[4]
Acremonium zeae	*Zea mays* L.	Pyrrocidines A, B	Fungi	[4]
Cryptosporiopsis quercina	*Tripterigeum wilfordii*	Cryptocandin Cryptocin	*C. albicans Trichophyton* spp. *Botrytis cinerea Pryriaria oryzae*	[4]
Alternaria sp. *Nigrospora oryzae Papulospora* sp.	8 types of medicinal plants found in 3 different regions of western India	Extract from fermented juice	*C. albicans*	[4]
Chaetomium globosum	Lead of *Ginkgo biloba*	Chaetoglobosin A and C	*Mucor miehei*	[4, 6]
Talaromyces sp.	Mangroves	7-epiaustdiol Stemphyperylenol Secalonic A	*P. aeruginosa Sarcina ventriculi*	[4, 7]

Table 1.
Antibiotic effects of some endogenous fungal species.

No.	Endophytic fungi	Isolation source	Ref.
1	*Pestalotiopsis microspora*	*Taxus wallichiana*	[8]
2	*Pestalotiopsis guepini*	*Wollemia nobilis*	[8]
3	*Seimatoantlerium tepuiense*	*Maguireothamnus sprciosus*	[8]

Table 2.
Some endophytic fungi producing paclitaxel.

1.6.3 Endophytic fungi that are active against insects

Many studies have demonstrated the importance of endophytic fungi in the production of insect repellents, which have many implications for crop protection and increase in agricultural yields **Table 3**.

Endophytic fungi	Isolation source	Compounds	Activities	Ref.
Nodulisporium sp.	*Bontia daphnoides*	Nodulisporic	Insecticide, against larvae of green flies, flies, etc	[4]
Muscodor vitigenus	*Paullina paullinioides*	Napthalene	Except for bed bugs	[4]

Table 3.
Anti-insect effects of some endophytic fungal species.

Endophytic fungi	Isolation source	Metabolites	Biological impact	Ref.
Pestalotiopsis microspora	*Terminalia morobensis*	Pestacine Isopestacine	Strong anti-oxidant Anti-fungal	[9]
Cephalosporium sp. *Microsphaeropsis olivacea*	*Trachelospermum jasminoides Pilgerodendron uviferum*	Graphislactone A	Stronger antioxidant than BHT and ascorbic acid	[10]
Xylaria sp.	*Ginkgo biloba*	Phenolic, flavonoid	Strong anti-oxidant	[10]
Phyllosticta sp.	*Guazuma tomentosa*	Metabolites	Strong anti-oxidant	[10]

Table 4.
Antioxidant effects of some endophytic fungal species.

Endophytic fungi	Isolation source	Metabolites	Biological impact	Ref.
Pseudomassaria sp.	Plants in the African Rainforest	Nonpeptidal (L-783,281)	Lowers blood sugar with a mechanism similar to insulin but taken orally	[4]
Fusarium subglutinans	*Trypterygium wifordii*	Subglutinol A Subglutinol B	Decreased B and T lymphocytes (immunosuppression)	[4]
Penicillium sp.	*Limonium tubiflorum*	NF-B inhibitor	Reduce the incidence of cancer	[4]
Pestalotiopsis microspora	*Torreya taxifolia*	Acid torreyanic	Anti-cancer agent	[4]
Fusarium solani	*Camptotheca acuminata*	Camptothecin	Anti-cancer compounds	[4]
Curvularia lunata	*Niphates olemda*	Cytoskyrin	Antimicrobial Potential agent for cancer treatment	[4]
Fusarium sp.	*Kandelia candel*	New isoflavone	Inhibits the growth of Hep-2 and Hep G2 cancer cells	[4]
Phomopsis sp.	*Musa acuminata*	Hexaketide γ-lacton Oblongolides Z	Anti-herpes simplex virus type 1	[4]
Phomopsis sp.	*Excoecaria agallocha*	Phomopsis-H76 A, B and C	Formation of vessels in the sub-intestinal vasculature	[4]

Table 5.
Endogenous fungi producing other biologically active substances.

1.6.4 Endogenous fungi with antioxidant activity

Antioxidants are substances that react with free radicals generated during oxidation, thus preventing or slowing down this process. Antioxidants prevent and treat diseases such as cancer, cardiovascular (atherosclerosis, hypertension, ischemia), diabetes, neurodegenerative diseases (Parkinson's disease), arthritis and aging, etc. Endophytic fungi in higher plants are a source of many new antioxidant active substances **Table 4** [11, 12].

1.6.5 Endogenous fungi produce other biologically active substances

Endogenous fungi are also known as a source of many other biological metabolites, such as anti-inflammatory, antidiabetic, hypoglycemic, immunosuppressive, etc. used to prevent rejection in organ transplants and can be used to treat autoimmune diseases such as rheumatoid arthritis, insulin dependent diabetes also known from endogenous fungi.

Endophytic fungi of the genus *Chaetomium* produce chaetoglobosin, which is a cytochalasin analog that inhibits actin polymerization. The genus *Chaetomium* is able to produce cytostatic metabolites including chaetomin, chaetoglobosin A, C, D and G, chaetoquadrin, oxaspirodion, chaetospiron, orsellide and chaetocyclinon **Table 5**.

2. Introduction of the endophytic *Aspergillus*

Aspergillus is a large genus of fungi and one of the most studied, according to Thom and Church in 1918, the genus is divided into 18 groups, 132 species and 18 orders. They account for a large proportion in the natural environment and are easy to culture in the laboratory. The economic importance of several species has attracted much research on *Aspergillus*. Furthermore, this common genus of fungi is involved in many industrial processes including production of enzymes (such as amylase), chemicals (such as citric acid) and food (such as soy sauce) are one of the plant endogenous fungal genera known for its ability to produce many biologically active substances with many practical applications [13].

2.1 Classify

Aspergillus was first classified in 1809 by the Italian priest and biologist Micheli by observation under a microscope. Nowaday, *"Aspergillum"* is also the name for asexual spores that form the common structure for all species of the genus *Aspergillus*; about a third of species in the genus have a sexual reproduction stage. According to the classification:

Kingdom: Fungi
Division: Ascomycota
Class: Eurotiomycetes
Order: Eurotiales
Family: Trichocomaceae
Genus: *Aspergillus*

2.2 Characteristics of *Aspergillus*

The characteristics of color (black, brown, yellow, red, white, blue, etc), growth rate, edge of mushroom cluster and surface texture of mushroom cluster vary

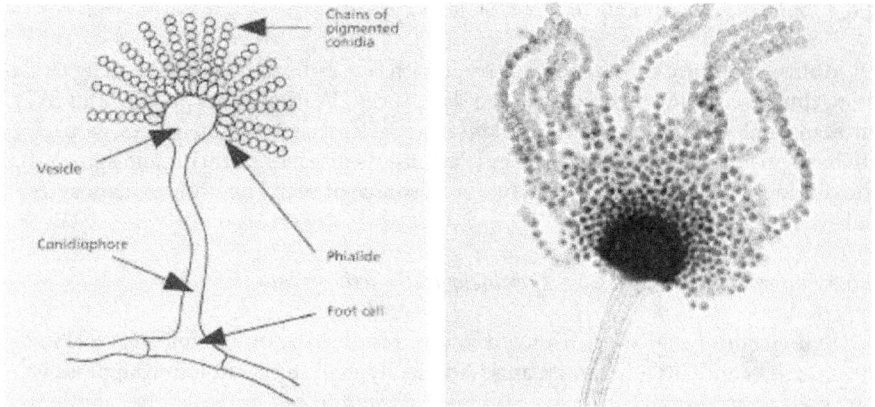

Figure 2.
Structure of the asexual reproductive organs of Aspergillus (source: Doddamani, 2012).

depending on species and growing conditions. The mycelium of *Aspergillus* belongs to the group of colorless, segmented, branched hyphae that can produce many enzymes and some toxins. *Aspergillus* mycelium is vigorous and produces many spores attached to a long vertical spore stalk, which grows from a special cell located in the trophic fiber called a foot cell [14, 15].

The spore-bearing head includes the spores: spore, flask, vesicle and spore stalk. The properties of each component vary from species to species and are characteristics that help identify species. Most species have the same shape, size, and color of spore-bearing heads as the cluster **Figure 2**.

Some common *Aspergillus* species: *A. aculeatus, A. candidus, A. flavus, A. foetidus, A. fumigatus, A.terreus, A. lentus, A. nidulans, A. niger, A. oryzae,* etc.

2.3 Ecological characteristics and distribution

Aspergillus is very aerobic, found in most oxygen-rich environments, usually growing on the surface of a substrate. Normally, fungi grow on carbon-rich substrates such as monosaccharides (glucose) and polysaccharides (amylase).

Aspergillus is widely distributed in the environment and can grow almost anywhere, especially in places with high humidity. *Aspergillus* grows by saprophyte on decaying plants, compost, and humus. Most *Aspergillus* species can live on a variety of substrates such as feces, human tissues, and ancient parchment. *Aspergillus* species grow well and produce many spores at a temperature of about 23–26°C. However, there are some species such as *A. terreus, Acronurus carneus, A. jcheri* which thrives at 35°C or *A. fumigatus* which grows well at 45°C even up to 50°C. Temperature also affects the shape of the attachment spore tip. *A. janus* species produces two different types of attached spore heads, the ratio of these two forms is affected by temperature, at 18–20°C, most of the spore heads are white, clubhead-shaped, and long spore stalks; but at 30°C, most of the spore heads are dark green, spherical, and short spore stalks.

2.4 Some biologically active substances produced by endophytic *Aspergillus*

Apergillus sp. has the ability to produce many biological active substances such as antibacterial, antiviral, cytotoxic and antioxidant activities, etc. **Table 6**.

Endophytic fungi	Isolation source	Metabolites	Biological impact	Ref.
A. flavus	Solanum lycopersicum L.	Chlorophyll, flavonoids, carbohydrates, phenolics, total proteins contents	To improve the growth and the secondary metabolites contents of tomato	[16]
A. flavus	Sonneratia alba	Kojic acid	Antibacterial (Staphylococcus aureus, Escherichia coli)	[17]
A. flavus IBRL-C8	Cassia siamea Lamk	The ethyl acetate extract	Antifungal (C. albican)	[18]
A.flavus	Viscum album	Lectin	Anti-cancer MCF7	[19]
A. flavus (SS03)	Moringa oleifera Lam.	Fenaclon, (R)(-) 14 methyl-8-hexadecyn-1-ol, Trans-β-farnesene (E)-β-farnesene, 9-Octcadecene,1,1, Dimethoxy	Antibacteria (S. aureus, Bacillus, Candida tropicalis)	[20]
A. flavus Nigrospora sphaerica	Tropical Tree Species of India, Tectona	Duroquinone, Adamantine derivative, Dodecanoic acid, tetradecanoic acid, pentadecanoic acid and Myristic acid	Insecticidal	[21]
A. flavus A. niger	Acacia nilotica	The ethyl acetate extract	Antifungal (Pythiummyriotylum, Rizoctoniasolani)	[22]
A. carbonarius A. niger	Zea mays Arachis hypogea	The ethyl acetate extract	Anticancer Promoted plant growth	[23]
A.niger CSR3	Cannabis sativa	Gibberellins, indoleacetic acid	Antibacteria	[24]
A.fumigatus	R7	Linoleic acid	Antimicrobial and cytotoxic activities	[25]
A.fumigatus	riparian plants Myricaria laxiflora	(Z)-N-(4-hydroxystyryl) formamide (NFA)	Improves drought resistance in rice as an antioxidant	[26]
A.fumigatus	Cocos nucifera	Flavonoid, terpenoid and saponin	Antibacteria (Pseudomonas aeruginosa, E. coli, Bacillus subtilis, S. aureus)	[27]
A.fumigatus	Copaifera multijuga	The compounds into the fermentation broth under specific culture conditions	Antibacteria (Mycobacterium tuberculosis H37Rv strain (ATCC 27294))	[28]
A.fumigatiaffinis	Tribulus terestris	A new antibacterial polyketide (-) palitantin	Antibacteria (Enterococcus faecalis UW 2689, Streptococcus pneumonia)	[29]
A. terreus KC 582297	The seaweed	The ethyl acetate extract	Antimicrobial	[30]

Endophytic fungi	Isolation source	Metabolites	Biological impact	Ref.
A.terreus (JAS-2)	*Achyranthus aspera*	The ethyl acetate extract	Antibacterial, antifungal and anti-oxidant	[31]
A. terreus	*Ambrosia ambrosoides*	Terrequinon A Terrefuranon Na-Acetyl aszonalemin	Anti-cancer	[32]
A. terreus	*Brickellia* sp.	Dehydrocurvularin 11-methoxycurvularin 11-hydroxycurvularin		
A.japonicus EuR-26	*Euphorbia indica L.*	Improved plant biomass and other growth features under high temperature stress (40∘C)	Modulate host plants growth under heat stress	[33]
A.nomius EF8-RSM	*Aloe vera* Western Ghats of Karnataka India	The ethyl acetate extract	Increases biomass production, increases synthesize different enzymes	[34]
A. iizukae	*Silybum marianum*	Silymarin (Silybin A (1), silybin B (2), and isosilybin A (3))		[35]
A.aculeatus	*Rosa damascena* Mill.	Secalonic acid	Anticancer (TNBC) cells.	[36]
A. tamarii	*Opuntia ficus-indica* Mill	The ethyl acetate extract	Against *Aedes aegypti and Culex quinquefasciatus*	[37]
A.clavatonanicus MJ31	*Mirabilis jalapa L.*	Seven antibiotics	Antimicrobial (*B. subtilis, Micrococccus luteus, S. aureus*)	[38]
A.aculeatinus Tax-6	Taxus yew barks	Taxol	Antitumor	[39]
Aspergillus sp.	*Ficus carica*	The ethyl acetate extract	Antimicrobial (*P. aeruginosa*)	[40]

Table 6.
Endophytic Apergillus sp. producing other biologically active substances.

2.5 Isolation of biologically active endophytic fungus *Aspergillus*

In many plants, the microflora is entirely endophytic fungal. This suggests that endophytic fungi may have a more favorable biological interaction than endophytic bacteria with respect to host plants. The strains of endogenous fungi with active substances are very diverse in both morphology and reproduction, and some strains have very special forms of reproduction.

Identification of *Aspergillus* strains by ITS gene sequencing method and searching on BLAST SEARCH gave similar results as the morphological method, contributing to the confirmation of strains with high biological activity which the subject isolated was *Aspergillus* [16].

Aspergillus spiecies isolated from plants have been shown to be able to produce many secondary metabolites with valuable biological effects such as anticancer, antiviral, antimicrobial compounds. Strains of *Aspergillus* isolated from galangal, turmeric, tangerine, and kumquat plants could produce metabolites with high activity against *S. aureus* and *MRSA*.

Conditions affecting the biological activity of the endophytic fungus Aspergillus

- pH: Importantly affects the growth, metabolism of fungi, enzyme activity, intermediate products, dissociation, dissolution, etc., thus affecting the biosynthesis of active ingredients antibacterial of fungi.

- Temperature: Like other microorganisms, the temperature of the environment also greatly affects the growth and development of fungi.

- Oxygen concentration: Oxygen concentration is very important and necessary for the survival and growth of aerobic microorganisms.

- Carbon source-nitrogen source: The choice of carbon and nitrogen sources greatly affects the activity of secondary substances. Different carbon sources such as dextrose, lactose, sucrose, fructose, starch, molasses and glycerol are believed to be suitable carbon sources for metabolism in various fungi. Organic and inorganic nitrogen sources such as $NaNO_3$, yeast extract, meat extract and soybean meal, NH_4NO_3, $(NH_4)_2SO_4$, etc. can help increase biological activity in fungi [2, 41, 42].

All the optimization was performed based on % inhibition of bacterial growth when challenged with 10 µg/µl metabolite concentration. Among different media used, potato dextrose broth (PDB) and sabouraud's dextrose broth (SDB) proven to be better media for growth of fungus as well as metabolites production 1% yeast extract and 4% dextrose resulted in higher cell inhibition. Ethyl acetate served as good extracting solvent [19].

- Addition of vegetable oil to the environment: Vegetable oil can be used to supplement the carbon source during lovastatin production in *Aspergillus*. Palm oil and soybean oil significantly increased the biomass and lovastatin production of *A. terreus* [43, 44].

- Trace elements: Fe and Zn are necessary for the biosynthesis of some antibiotics. It is possible that these two minerals have a positive effect on the antibiotic biosynthesis of *Aspergillus*.

- Salt concentration: Salt concentration affects the antibacterial activity of *Aspergillus*. For example, strain *A. terreus* has strong antibiotic activity in the range of NaCl salt concentrations from 0 to 1%. When the salt concentration is above 1%, the antibiotic activity of this strain decreases and at a NaCl concentration of 6%, *A. terreus* strain is no longer capable of biologically active substances. *A. terreus* is not only of research interest in terms of antimicrobial activity, but it is well known for its ability to produce lovastatin. According to the study of Pawlak et al. in 2012 on optimal conditions for lovastatin production of *A. terreus,* the authors determined that ventilation is essential for aerobic biological response [45].

Endophytic fungal populations of the genus *Aspergillus* have been isolated from many plants and have been shown to produce a wide range of biologically active substances including antifungal, antibacterial, anticancer, etc. In addition to the characteristics of resistance to *MRSA* and *S. aureus* as published by many studies, some strains of *A. terreus* isolated from soil or from plants have a spectrum of effects on a number of other bacteria such as *E. coli, P. aeruginosa, Streptococcus faecalis*. This shows the potential to study antibacterial compounds of *Aspergillus*.

2.6 Determination of biological activity of secondary compounds

In the world, there are many studies on the role and application of biologically active substances produced by endogenous fungi. Some endophytic fungal strains have the ability to produce important antibiotics to prevent the invasion of pathogenic organisms to host plants, which are significant in the control of plant diseases and insect pests. Some endogenous fungi are able to synthesize biologically active substances used as anticancer drugs, produce tumor suppressor antibiotics, immunostimulants, and antioxidants, and have biological activities. These compounds mainly belong to the groups of alkaloids, steroids, flavonoids and terpenoid derivatives and other substances, etc. Endogenous fungi also perform a resistance mechanism against plant diseases by producing substance with antibacterial activity. Screening for antimicrobial compounds from endogenous fungi is a way to kill resistant bacteria in humans and plants. In addition, the natural metabolites of endogenous fungi also help to protect natural resources and meet the requirements of pharmaceutical production from plant origin by fermentation. Many biologically active substances are produced by endogenous fungi during growth and development. Finding and discovering those active ingredients is the goal that biopharmaceutical researchers are constantly reaching for.

2.7 The interaction between the host plant and the endophytic fungus *Aspergillus*

There is a complex relationship between endophytic fungi and host plants, the interaction between host and endophytic fungi can be endogenous or symbiotic depending on genetic predisposition, developmental stage, nutritional status and environmental factors.

Commensalism helps the endogenous fungi to survive by being supplied with nutrients without affecting the host plant. Mutual beneficial relationships of endophytic fungi and host plants through the provision of energy, nutrients, shelter as well as protection under environmental stress. On the other hand, endophytic fungi indirectly benefit from host plant growth by producing secondary metabolites that help host plants adapt to abiotic factors such as light, drought and stress such as herbivores, insect and nematode attacks or pathogens.

Schulz and Boyle in 2005, the authors proposed that the endophysis of endo-phytic fungal is a balanced antagonism between host and endophytic fungi, and provided endogenous virulence and protective capacity of the balanced host plant showed no significant symptoms.

Once the host-endophytic interactions become imbalanced, or disease in the host plant or host defense tissues kills the pathogenic endophytic fungi. Whether the interaction is balanced or unbalanced depends on the host-endophytic condition, virulence of the fungi, host defenses, toxicity, environment, and nutritional status and growth stages of the host plant and endophytic fungi.

Therefore, commensal relationships require a balance between the defense responses of the host plant and the nutritional requirements of the endophytic fungi. In agreement with the 2006 study by Kogel et al., endophytic fungi share structural similarities with pathogens and both have many similar virulence factors, such as production of Metabolites and exoenzymes are required to infect and colonize the host plant, so endophytic fungi are subject to self-recognition, the host plant can respond to defensive responses as a disease agent. In addition, the cell wall of endogenous fungi is often associated with the production of macromolecular compounds in plants. Thus, endogenous fungi avoid or overcome nonspecific resistance to invasion by programming the invading cells to harbor pathogenic structures and to maintain integrity in the host cell for a long time [4, 46].

3. Conclusion

Isolation of endophytic fungi from medicinal and other plants may result in methods to produce biologically active agents for biological utilization on a large commercial scale as they are easily cultured in laboratory and fermentor instead of harvesting plants and affecting the environmental biodiversity.

Author details

Vo Thi Ngoc My* and Nguyen Van Thanh
Nguyen Tat Thanh University, Ho Chi Minh City, Vietnam

*Address all correspondence to: vtnmy@ntt.edu.vn

IntechOpen

References

[1] Tan RX, Zou WX (2001), "Endophytes: a rich source of functional metabolites", Natural Product Reports, 18, pp. 448 – 459.

[2] Hawksworth DL, Rossman AY (1997), "Where Are All the Undescribed Fungi?", Phytopathology, 87 (9), pp. 888-891.

[3] Strobel G và Daisy B (2003), "Bioprospecting for microbial endophytes and their natural products", *Department of Plant Sciences,* Microbiology and molecular biology, 67(4), pp. 491-502.

[4] Selim KA, El-Beih AA, AbdEl-Rahman TM, El-Diwany AI (2012), "Biology of endophytic fungi", Current Research in Environmental and Applied Mycology, 2(1), pp. 31-82.

[5] Tuntiwachwuttikul P, Taechowisan T (2008), *"Secondary metabolites from Streptomyces sp.", SUC1.* Tetrahedron, 64, pp. 7583-7586.

[6] Castillo UF, Strobel G et al. (2000), "Munumbicins, Wide-spectrums antibiotics produced by *Streptomyces* NRRL 30562, endophytic on *Kennedia nigriscans*", Microbiology, 148, pp. 2675 – 2685.

[7] Raviraja NS, Maria GL, Sridhar KR (2006), "Antimicrobial evaluation of endophytic fungi inhabiting medicinal plants of the Western Ghats of India", Engineering In Life Sciences, 6 (5), pp. 515-520.

[8] Tsui TY, Brow GD (1996), "Chromones and Chromanones from Baeckea frutescens", Phytochemistry, 43 (4), pp. 871-876.

[9] Davis EC , Franklin JB , Shaw AJ and Rytas V (2003), "Endophytic Xylaria (Xylariaceae) among liverworts and angiosperms: phylogenetics, distribution, and symbiosis", American Journal of Botany, 90(11), pp. 1661-1667.

[10] Srinivasan K, Jagadish LK (2010), "Antioxidant activity of endophytic fungus *Phyllosticta* sp. isolated from *Guazuma tomentosa*", Journal of Phytology 2 (6), pp. 37-41.

[11] Arora DS, Chandra P (2010), "Assay of antioxidant potential of two *Aspergillus* isolates by different methods under various physio-chemical conditions", Brazilian Journal of Microbiology, 41, pp. 765-777.

[12] Seema D, Sandeep K (2012), "Antioxidant activity of fungal endophytes isolated from salvador oleoides decne", International Journal of Pharmacy and Pharmaceutical Sciences, 4 (2), 380-385.

[13] Thom C, Church MB (1918). "*Aspergillus fumigatus, A. nidulans, A. terreus n.* sp. and their allies", American Journal of Botany, 5 (2), pp. 84-104.

[14] Masayuki M, Katsuya G, Bennett JW (2010), *"Aspergillus*: Molecular biology and genomics", *Caister Academic Press,* pp. 220-238.

[15] Mariana (2013), "Fungal infections *Aspergillus terreus*", *Leading International Fungal Education*, pp. 174-176.

[16] Fatma Abdel-Motaal, Noha Kamel, Soad El-Zayat, Mohamed Abou-Ellail (2020), "Early blight suppression and plant growth promotion potential of the endophyte *Aspergillus flavus* in tomato plant", Annals of Agricultural Sciences, 65, pp. 117-123.

[17] Antonius R. B. Ola, Christina A. P. Soa, YosephSugi, Theo Da Cunha, Henderiana L. L. Belli, Herianus J. D. Lalel (2020), "Antimicrobial metabolite from the endophytic fungi *Aspergillus*

flavus isolated from *Sonneratia alba*, a mangrove plant of Timor- Indonesia", Rasayan J. Chem., 13(1), pp. 377-381.

[18] Darah Ibrahim, Nurhaida, Lim Sheh Hong (2018), "Anti-candidal activity of *Aspergillus flavus* IBRL-C8, an endophytic fungus isolated from *Cassia siamea* Lamk leaf", *Journal of Applied Pharmaceutical Science*, Vol. 8(02), pp. 083-087.

[19] Sadananda TS, Govindappa M, Ramachandra YL, Chandrappa CP, Umashankar T (2016), "*In Vitro* Apoptotic Activity of Endophytic Fungal Lectin Isolated from Endophyte, *Aspergillus flavus of Viscum album* on Human Breast Adenocarcinoma Cell Line (MCF-7)", Metabolomics 6: 162.

[20] RAJESWARI S., UMAMAHESWARI S., D. ARVIND PRASANTH, RAJAMANIKANDAN K. C. P. (2016), "Bioactive potential of endophytic fungi *Aspergillus flavus* (SS03) against clinical isolates", *International Journal of Pharmacy and Pharmaceutical Sciences*, Vol. 8 (9).

[21] N. Senthilkumar, S. Murugesan, D. Suresh Babu (2014), "Metabolite Profiling of the Extracts of Endophytic Fungi of Entomopathogenic Significance, *Aspergillus flavus* and *Nigrospora sphaerica* Isolated from Tropical Tree Species of India", *Tectona grandis* L., *Journal of Agriculture and Life Sciences,* Vol. 1 (1).

[22] Dipali B. Tribhuvan, Aparna S.Tawre (2019), "Antifungal activity of endophyte *Aspergillus flavus* isolated from *Acacia nilotica*", *International Journal of Current Research*, Vol. 11, (07), pp. 5139-5140.

[23] EDWIN RENE PALENCIA (2012), "Endophytic associations of species in the *Aspergillus section nigri* with maize (*Zea mays*) and peanut (*Arachis hypogea*) hosts and their mycotoxins, *Under the Direction of Charles W. Bacon.*

[24] Lubna, Sajjad Asaf, Muhammad Hamayun, Humaira Gul, In-Jung Lee,

Anwar Hussain (2018), "*Aspergillus niger* CSR3 regulates plant endogenous hormones and secondary metabolites by producing gibberellins and indoleacetic acid", Journal of Plant Interactions, 13 (1), pp. 100-111.

[25] Mohamed Shaaban, Hamdi Nasr, Amal Z. Hassan, Mohsen S. Asker (2013), "Bioactive secondary metabolites from endophytic *Aspergillus fumigatus*: Structural elucidation and bioactivity studies", *Rev. Latinoamer. Quím.* 41/1.

[26] Wanggege Qin, Chengxiong Liu, Wei Jiang, Yanhong Xue, Guangxi Wang and Shiping Liu (2019), "A coumarin analogue NFA from endophytic *Aspergillus fumigatus* improves drought resistance in rice as an antioxidant", BMC Microbiology 19:50.

[27] Guerrero, J. J. G., Imperial, J. T., General, M. A., ArenaE. A. A., Bernal, M. B. R. (2020), "Antibacterial activities of secondary metabolites of endophytic *Aspergillus fumigatus*, *Aspergillus* sp. and *Diaporthe* sp. isolated from medicinal plants", Österr. Z. Pilzk, 28: 53-61.

[28] Silva, E.M.S.; Silva, I.R.; Ogusku, M.M.; Carvalho, C.M.; Maki, C.S.; Procópio, R.E.L. (2018), "Metabolites from endophytic *Aspergillus fumigatus* and their in vitro effect against the causal agent of tuberculosis", Acta Amazonica 48: 63-69.

[29] Antonius R. B. Olaa,b, Bibiana D Tawoa, Henderiana L. L Bellib, Peter Prokschc, Dhana Tommyc, Euis Holisotan Hakimd (2018), "A New Antibacterial Polyketide from the Endophytic Fungi Aspergillus fumigatiaffinis", *Natural Product Communications,* Vol. 13 (12), pp. 1573-1574.

[30] Suja Mathan, Vasuki Subramanian and Sajitha Nagamony (2013), "Optimization and antimicrobial metabolite production from endophytic

fungi *Aspergillus terreus* KC 582297",
Euro. J. Exp. Bio., 3(4):138-144.

[31] Goutam J, Singh S, Kharwar RN, Ramaraj V (2016), "*In vitro* Potential of Endophytic Fungus *Aspergillus terreus* (JAS-2) Associated with *Achyranthus aspera* and Study on its Culture Conditions", Biol Med (Aligarh) 8: 349.

[32] Hedayati MT, Pasqualotto, Warn; Bowyer, Denning (2007),"Aspergillus flavus: human pathogen, allergen and mycotoxin producer"Aspergillus flavus: human pathogen, allergen and mycotoxin producer", Journal of Medical Microbiology, 153 (6), pp. 1677-1692.

[33] Ismail, Muhammad Hamayun, Anwar Hussain, Amjad Iqbal, Sumera Afzal Khan, In-Jung Lee (2018), "Endophytic Fungus Aspergillus japonicus Mediates Host Plant Growth under Normal and Heat Stress Conditions", *BioMed Research International*, Vol. 2018.

[34] Rohit Shankar Mane and Ankala Basappa Vedamurthy (2020), "Physiology of Endophytic *Aspergillus nomius* EF8-RSM Isolated from *Aloe vera* Western Ghats of Karnataka India", Asian Journal of Applied Sciences, 13 (1), pp. 32-39.

[35] El-Elimat T., Raja H.A., Graf T.N., Faeth S.H., Cech N.B., and Oberlies N.H. (2014), "Flavonolignans from *Aspergillus iizukae*, a Fungal Endophyte of Milk Thistle (*Silybum marianum*)", Journal of Natural Products, 77, pp.193-199.

[36] Sadaqat Farooq, Arem Qayum, Yedukondalu Nalli, Gianluigi Lauro, Maria Giovanna Chini (2020), "Discovery of a Secalonic Acid Derivative from *Aspergillus aculeatus*, an Endophyte of *Rosa damascena* Mill.", *Triggers Apoptosis in MDAMB-231 Triple Negative Breast Cancer Cells ACS Omega*, 5, pp. 24296–24310.

[37] Kannan Baskar, Ragavendran Chinnasamy, Karthika Pandy, Manigandan Venkatesan (2020), "Larvicidal and histopathology effect of endophytic fungal extracts of *Aspergillus tamarii* against *Aedes aegypti* and *Culex quinquefasciatus*", Heliyon 6.

[38] Mishra VK, Passari AK, Chandra P, Leo VV, Kumar B, Uthandi S, et al. (2017), "Determination and production of antimicrobial compounds by *Aspergillus clavatonanicus* strain MJ31, an endophytic fungus from *Mirabilis jalapa L.* using UPLC-ESI-MS/MS and TD-GC-MS analysis", PLoS ONE *12(10)*.

[39] Weichuan Qiao, Tianhao T ang, Fei Ling (2020), "Comparative transcriptome analysis of a taxol-producing endophytic fungus, *Aspergillus aculeatinus* Tax-6, and its mutant strain", Scientific Reports, 10:10558.

[40] D. Prabavathy, C. Valli Nachiyar (2011), "Screening and characterisation of antimicrobial compound from endophytic *Aspergillus* sp. isolated from *Ficus carica*", *Journal of Pharmacy Research*, Vol.4 (6).

[41] Gao, (2007),"Effects of carbon concentration and carbon to nitrogen ratio on the growth and sporulation of several biocontrol fungi", Mycological Research, 111, pp. 87-92.

[42] Pranay J, Ram KP (2010), "Effect of different carbon and nitrogen sources on *Aspergillus terreus* antimicrobial metabolite production", International Journal of Pharmaceutical Sciences Review and Research, 5(3), pp. 72-76.

[43] Sripalakit P, Riunkesorn J, Saraphanchotiwitthaya A (2011), "Utilisation of vegetable oils in the production of lovastatin by *Aspergillus terreus* ATCC 20542 in submerged cultivation", Maejo International Journal of Science and Technology, 5(02), pp. 231-240.

[44] Zaehle C, Gressler M, Shelest E, Geib E, Hertweck C, Brock M (2012), "Microbial transformation of the sesquiterpene lactone tagitinin C by the fungus *Aspergillus terreus*", *Journal of Industrial Microbiology and Biotechnology*, pp. 1214-1216.

[45] Pawlak M, Marcin B (2012), "Kinetic modelling of lovastatin biosynthesis by *Aspergillus terreus* cultivated on lactose and glycerol as carbon sources", Chemical And Process Engineering, 33 (4), pp. 651-665.

[46] Huawei Zhang, Yifei Tang1, Chuanfen Ruan1 and Xuelian Bai (2016), "Bioactive Secondary Metabolites from the Endophytic *Aspergillus* Genus", Rec. Nat. Prod. 10:1, pp. 1-16.

Chapter 5

Carbon Storage of Some Rubber Trees (*Hevea brasiliensis*) Clones in HEVECAM's Plantations in South Cameroon

Menoh A. Ngon René, Tsoata Esaïe,

Tsouga Manga Milie Lionelle and Owona Ndongo Pierre-André

Abstract

The objective of this work was to estimate the quantity of carbon stored by four main clones of rubber tree cultivated in South Cameroon: GT 1, PB 217, PR 107 and RRIC 100. The forest inventory method was used to measure trees morphological parameters, the latter used to calculate carbon storage using the allometric equation of Wauters et al., (2008). The main morphological parameters measured were: leaf area index (LAI), circumference (C), diameter at breast height (DBH) and total tree height (h). Comparing the morphological parameters of clones two by two using a Dunn test, we observe significant differences in the circumference, the diameter and even very significant in the leaf area index, but not in the height. The clones GT 1, PR 107, PB 217, and RRIC 100 stored on average: 111.05 tC / ha, 150.18 tC / ha, 165.25 tC / ha, and 187.25 tC/ha respectively. A significant difference was established between the means of carbon storage of the clones GT 1 and PB 217 (p = 0.0488) on one hand and, that of the clones GT 1 and RRIC 100 (p = 0.0240), on the other hand. These results are an estimation of models, further research can be undertaken for exact measurements.

Keywords: carbon storage, rubber tree, clones, HEVECAM, Cameroon

1. Introduction

According to the IPCC (Intergovernmental Panel on Climate Change), since the industrial revolution, the concentrations of greenhouse gases (GHGs) in the atmosphere have only increased and carbon dioxide (CO_2) constitutes 76.7% of this increase [1]. Mitigating global warming is a major concern and inevitably constitutes one of the major challenges of the present century [2]. Negotiations during the conferences of the parties (COP) of the UNFCCC (United Nations Framework Convention on Climate Change), led in 2012 during the COP 18 in Doha, to an objective of reducing GHG emissions by at least 18% during the period 2013–2020, compared to the level of the 1990s [3]. During the COP 21, the President of the Republic of Cameroon (PRC) asserted that Cameroon ratified the UNFCCC in 1994 and pledged to reduce its GHG emissions by 32% by 2035 compared to 2010 [4]. Thus, reliable estimates of the quantities of carbon stored by the various plant formations are necessary in order to be able to count them in the reduction of GHGs. The purpose of quantifying carbon storage is therefore to evaluate efforts to reduce CO_2 emissions and fight against global warming [5, 6]. Studies for estimating the amounts of carbon stored has long been limited to natural forests neglecting the storage potential of planted forests and agricultural plantations [7]. However a study conducted by Makundi in African forest plantations revealed that, they stored nearly 40 million tonnes of CO_2 per year. Those tonnes of CO_2 could be credited with REDD + (reducing greenhouse gas emissions from deforestation and forest degradation, promoting conservation, forest management) in Africa [8]. In Cameroon, *Hevea brasiliensis* plantations are estimated at more than 52,000 ha planted and in full extension [9]. However, there would be very little research on the carbon storage potential of these plantations [10]. In addition, existing studies on the estimation of carbon storage by *Hevea brasiliensis* have not taken into account the clonal differences [10, 11]. So this study fills the gap in a neglected sector of research on carbon storage in Cameroon. The results of the latter could be used in the implementation of the reduction of GHG emissions due to REDD+ through the village plantations of *Hevea brasiliensis*, and allow Cameroon to meet its commitments to reduce GHG emissions. This research is based on the question of whether the amount of carbon stored by *Hevea brasiliensis* varies between clones. The general objective of this work was to estimate the amount of carbon stored by four cultivated clones of *Hevea brasiliensis*. Specifically, the study aimed to: determinate some morphological parameters of the clones cultivated at HEVECAM; estimate the aboveground biomass (AGB) and estimate the carbon stored by the different clones.

2. Material and methods

2.1 Material

The study was carried out in HEVECAM plantation located in the Southern region of Cameroon, Niété subdivision (2° 40'North, 10° 03' East) (**Figure 1**). Established in 1975, but the examined trees was 24 years. These plantations are the largest plantations of *Hevea brasiliensis* in Cameroon. It is also the government largest cultivation project of *Hevea brasiliensis* and the largest "development society" in South Cameroon [13]. HEVECAM was acquired by several groups, including Corrie Maccoll Limited, the American group to which it currently belongs.

The climate is equatorial of the Guinean type, marked by four seasons: a long dry season (from November to February with a peak in December); a small rainy season (from March to May with the peak in May); a short dry season (from June to mid-August); and a long rainy season (mid-August to November); The average annual rainfall is 3000 mm and the average annual temperature is 27° C. The relief is between the southern plateau and the coastal plain. The altitude varies between 20 and 300 m. The population is cosmopolitan, made up mostly of nationals from the far North, North and South West regions of Cameroon.

The HEVECAM plantation occupies approximately 22,000 ha, subdivided into 17 villages (numbered from V 1 to V 17). Each of these villages has blocks divided into plots. This research took place in V 10 and V 15. The choice of villages was based on the presence in the village and the suitable age of the clones retained in the study. Four clones was chosen for this study, because of their performance in this area: PB 217, GT 1, RRIC 100 and PR 107.

2.2 Methods

The estimation of the storage potential was made using the forest inventory method because it was the most suitable for this study [14]. This method consists in

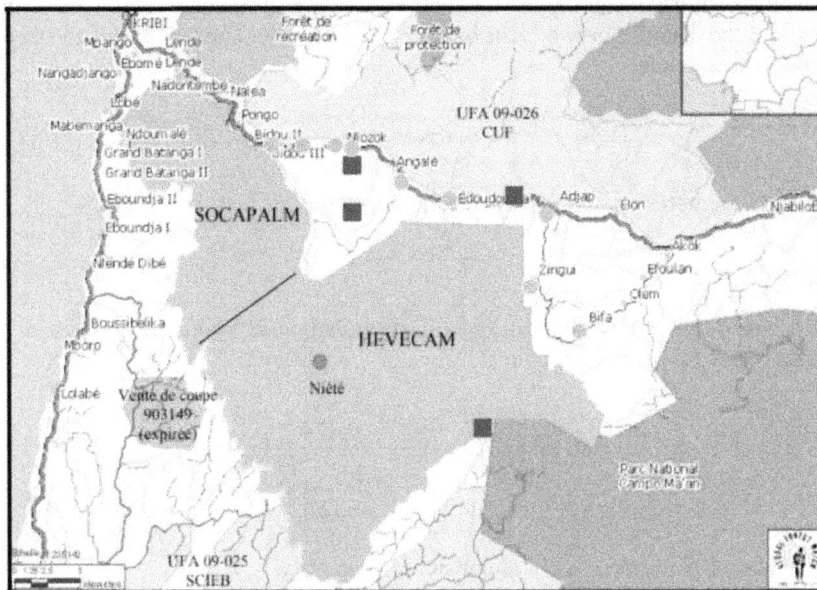

Figure 1.
Location of the HEVECAM plantation. Source [12].

recording the morphological parameters of standing trees and calculating the carbon stored using allometric equations relating the parameters to the carbon stored. Sampling was done according to the method proposed by [15]. Indeed, 4 temporary plots each measuring 25 m x 25 m were installed for the four clones to be studied (ie one plot per clone). Plots of this size are recommended, as they are more effective in a monoculture system such as *Hevea brasiliensis* plantations [15]. In addition, this size is recommended for trees with a diameter between 20 and 50 cm, as is the case in this study [15]. In each of these four plots, measurements were taken on 10 trees, so a total of 40 trees was used for the study. Ten trees were chosen because 8 to 10 trees are sufficient for taking measurements in a plot of this size 25 m x 25 m, and that this is fairly representative of a hectare of plantation [15].

Overall, there were 4 samples plots each with 45 trees planted in a row, 6 m between rows and 3 m between trees in the same row, for a density of 555 trees/ha. In all, 4 plots, data were collected on 40 randomly selected trees. The data collected were: the DHB at 1.5 m height because it had to be above the bled panel; the height h of the tree using a clinometer.

2.2.1 Diameter at breast height measurement

The diameter was obtained from the circumference measured at 1.5 m from the ground using a tape and a 1.5 m pole [16].

2.2.2 Height measurement

This is the total height of the tree from the foot to the terminal bud of the tree. It was measured using a Steren clinometer [17]. The operator stands at a distance (B) as close as possible to the estimated height of the tree. Then, through the dioptric viewfinder of the clinometer, he aims to turn at the top and then at the base of the tree. On each side, he notes the graduation to the right of the dial. This is the tangent of the angle of inclination expressed as a percentage (%). Let α be the angle of inclination with the foot of the tree, β the angle of inclination with the top of the tree and B the distance between the operator and the tree. The height h of the tree is given by the formula:

$$h = B(\tan\alpha + \tan\beta) \qquad (1)$$

h and B in meters, α and β in degrees.

2.2.3 Leaf area index determination

LAI was determined using the logistic regression model proposed by [18], express by the following formula:

$$\ln LAI = 1.225 + \frac{0.474}{1 + \left(\dfrac{\ln D^2}{6.327}\right)^{-21.48}} \qquad (2)$$

D is the DBH in cm; LAI is a dimensionless quantity.

2.2.4 Quantifying living aboveground biomass

The above – ground biomass (AGB) is the mass of the entire upper part of the plant which includes: the trunk, branches and leaves. It was calculated on tree scale using the allometric equation proposed by Dey et al. [19], then converted to the hectare using the planting density of the plantation.

$$AGB = 0,0202\, C^{2,249} \qquad (3)$$

(C) is the circumference at 1.5 m from the ground in cm. (AGB) the above-ground biomass in Kg/tree.

This model was preferred because it is suitable for our research for three main reasons: It is specific to *Hevea brasiliensis* and not generic to several species like the majority of models; in addition, the use of the circumference instead of the diameter reduces the errors that can occur in the calculation of the diameter; moreover, this model allows that by measuring the circumference at 1.50 m from the ground, one avoids the bleeding panel generally located at 1.30 m from the ground [19].

2.2.5 Calculation of the stored carbon by the different clones

The estimate of stored carbon was carried out according to one of the methods propose by [20]. In fact, we determine the average carbon stock per tree and by multiplying by the density, we get the carbon per hectare. To do this, we used the allometric equation specific to *Hevea brasiliensis* proposed by [20].

$$\ln CS = -5,147 + 2,392 \ln C \qquad (4)$$

C represents the circumference measured at 1.5 m the ground and CS the total carbon stored expressed in Kg/tree.

This model was preferred over the others for two main reasons: first, it is specific to the species *Hevea brasiliensis* and the trees used to establish this equation are practically the same age as those in the present study [20].

2.3 Statistical analyzes

The data were processed (ordered, classified and grouped) by the Excel software from which an input mask was obtained and later analyzed using R software. The main tests performed is the Dunn test. It made it possible to identify pairs of clones whose variables are significantly different [21].

3. Results

3.1 Determination of morphological parameters of clones

When comparing the morphological parameters of clones two by two using a Dunn test, we observe significant differences in the circumference, the diameter and even very significant in the leaf area index, but not in the height (**Table 1**).

		RRIC 100	PB 217	GT 1
DBH	GT 1	0,014*	0,032*	
LAI	GT 1	0,001**	0,010*	
	PR 107			0,467

Meaning of the codes: "" = significant; "**" = very significant.*

Table 1.
Dunn's test of morphological parameters.

3.2 Estimation of aboveground biomass and carbon storage

The diagram shows that the RRIC 100 clone store the most carbon, i.e., 187.11 tC /ha and GT 1 the less carbon (111.04 tC/ha) (**Figure 2**). By performing a

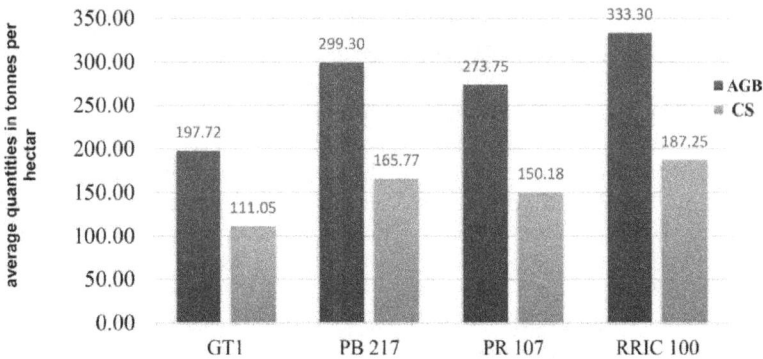

Figure 2.
Average quantities of above-ground biomass and carbon stored as a function of clones.

two-by-two comparison using a Dunn test (**Table 2**), a significant difference is observed (0.001 < p < 0.05) between two pairs of clones: PB 217 and PR 107 (p-value = 0.0488); GT 1 and RRIC 100 (p-value = 0.0240). The other comparisons show non-significant differences.

		p-value	p-value Signification
GT 1	PB 217	0,0488	*
GT 1	PR 107	0,0961	Ns
GT 1	RRIC 100	0,0240	*
PB 217	PR 107	0,760	Ns
PB 217	RRIC 100	0,774	Ns
PR 107	RRIC 100	0,553	Ns

Meaning of the codes: "" = significant; Ns = not significant.*

Table 2.
Dunn's test of the carbon stored by the different clones.

4. Discussion and conclusion

4.1 Discussion

4.1.1 Morphological parameters of the studied clones

In this research it was established that there is a very significant difference between the leaf area indices of the studied clones. The RRIC 100 clone has the highest LAI, ie 4.98, a value different from that obtained by [22], who obtained for the same clone, a leaf area index of 3.37. The difference between these values could be explained by the differences in the age between the trees, 20 years for [22] and 24 years for the present study. In addition, [22] used a direct measurement method using a leaf area meter. Concerning circumferences, a study by [23] on the GT 1 clone obtained an average of 100 cm, which is different from the obtained 75 cm for the GT 1 clone in this study. Once again, the age difference between the trees of the two studies could justify the difference in the results, as the circumference increases with the age of the trees. Trees in the study of [23] were slightly older (33 years) than the trees in the present study. On the other hand, [6] obtained an average DBH of 24.9 ± 0.7 for 25-year-old trees. Although the study [6] does not give details on the clones studied, it is noted that this diameter is substantially equal to the mean diameter of the GT 1 clone (24.01 ± 5.07) of the present study.

4.1.2 Quantity of aboveground biomass and stored carbon

AGB is between 197 and 333 tC / ha, and stored carbon between 111.05 and 187.25 tC / ha. The RRIC 100 clone stored more carbon than the other four clones and the difference is significant or even highly significant with the other clones [24]. Obtained 214 tC/ha, a value closer to what we obtained in this work (333 tC/ha). The trees in this study are 24 years old and in the studies of [24] they were 31 years. It is therefore understandable why the results we obtained are closer to those in [24], in view of the approximation of age.

Compared to other ecosystems, rubber plantations can store more carbon than secondary forests of the same age, which the storage capacity varies with the age between 91.75–256.5 tC/ha, according to [25].

4.2 Conclusion

The question behind this work was to know whether the carbon storage potential of the *Hevea brasiliensis* species varies among clones. To answer this question, it was necessary on one hand to determine the morphological parameters related to carbon storage and on the other hand to quantify the biomass and then the storage potential of 4 *Hevea brasiliensis* clones. From the four clones studied, it was established that there is a significant difference in the means of carbon stored between the GT 1 and RRIC 100 clones on the one hand and between GT 1 and PB 217 on the other hand. The clone RRIC 100 exhibits the greatest average carbon stored (187.11 tC/ha) These results are an estimation of models, further research can be undertaken for exact measurements.

Author details

Menoh A. Ngon René[1,2*], Tsoata Esaïe[1], Tsouga Manga Milie Lionelle[1,2] and Owona Ndongo Pierre-André[2]

1 Department of Plant Biology, University of Yaoundé I, Cameroon

2 Institute of Agricultural Research for Development (IRAD), Cameroon

*Address all correspondence to: menohr@yahoo.com

IntechOpen

References

[1] IPCC (Intergovernmental Panel on Climate Change), 2013. Summary for Policymakers, Climate Change: The Scientific Evidence. Contribution of Working Group I to the Assessment Report of the Intergovernmental Panel on Climate Change (IPCC) Cambridge University Press, Cambridge, United Kingdom and New York United States of America, 204 p.

[2] IPCC (Intergovernmental Panel on Climate Change), 2008. 2007 review of changes climate: Synthesis report contribution of working groups I, II and III to the fourth assessment report. IPCC, Geneva, Switzerland, pp. 14-18.

[3] UNFCCC (United Nations Framework Convention on Climate Change), 2012. Doha amendment to the Kyoto Protocol, Doha, Qatar, 2012, 6 p.

[4] PRC (The President of the republic of Cameroon) 2015. The speech of the President of the Republic of Cameroon at COP 21. Paris, November 30, 2015. Available on www.prc.cm, consulted on 09/30/2019, 2p.

[5] Biplan B., Arun J. N., Ashesh K. D., 2016. Managing rubber plantations for advancing climate change mitigation strategy. Research communications. *Curr., Sci.*, 110 (10) : 2015 – 2019.

[6] Liu C., Pang J., Rudbeck J. M., Lü X., Tang J., 2017. Carbon stocks across a fifty-year chronosequence of rubber plantations in tropical China. *For.* ,8 (209) : 14p.

[7] Whittinghill L. J, Rowe D. B., Schutzki R., Cregg B. M., 2014. Quantifying carbon sequestration of various green roof and ornamental landscape systems. Landsc. Urb. Plan, 123: 41-48.

[8] Makundi R. W., 2014. Perspectives de REDD+ dans les plantations forestières

africaines in *Afric*. For. Forum, 2(5): 28 p.

[9] Bahri-Domon Y., 2016. Agribusiness *business in Cameroun*, november 2016, N° 45, pp. 24.

[10] Egbe A. E., Tabot P. T., Fonge B. A., Bechem E., 2012. Simulation of the impacts of three management regimes on carbon sinks in rubber and oil palm plantation ecosystems of South-Western Cameroon. J. Ecol. Nat. Envir. 4(6) : 154 – 162.

[11] Njukeng N. J., Ehabe E. E., 2016. above ground biomass and carbon stock in some perennial crop-based agroforestry systems in the humid forest zone of Cameroon. ACRI., 5(1) : 1-13.

[12] Global Forest Watch (GFW). 2005. Atlas forestier interactif du Cameroun (version 1.0). Yaoundé : Ministère de l'Environnement et des Forêts; Yaoundé: Global Forest Watch Cameroon; Washington, DC: World Resources Institute

[13] Ngotta B. J. B., Dibong S. D., Taffouo V. D., Ondoua J. M., Bilong P., 2015. Niveau de parasitisme des hévéas par les *Loranthaceae* dans la région du Sud-Ouest Cameroun. J. Appl. Biosci., 96: 9055 – 9062.

[14] Gibbs H. K., Brown S., Niles J. O., Foley J. A., 2007. Monitoring and estimating tropical Forest carbon stocks: making REDD a reality. Environ. Res. Lett., 2, 13p.

[15] Pearson T., Walker S., Brown S., 2005. Sourcebook for land use, land-use change and forestry projects. *Winrock International*, Montreal, Canada, 56 p.

[16] Kurniatun H., Sitompul S.M., Van Noordwijk M., Palm C., 2001. Methods for sampling carbon stocks above and below ground. *ICRA*, Bogor, Indonesia, 23p.

[17] Weyerhaeuser H., Tennigkeit T., 2000. Forest inventory and monitoring manual. International Centre for Research in Agroforestry (ICRAF). Chiang Mai, Thailand, 30p.

[18] Chaturvedi R. K., Shivam Singh, Hema Singh, Raghubanshi A. S., 2017. Assessment of allometric models for leaf area index estimation of *Tectona grandis*. Trop. Pl. Res., 4 (2): 274 – 285.

[19] Dey S. K, Chaudhuri D., Vinod K. K., Pothen J. and Sethuraj M. R., 1996. Estimation of biomass in *Hevea* clones by regression method: 2. Relation of girth and biomass for mature trees of clone RRIM 600. *Indian J. of Nat. Rubber Res.*, 9(1) : 40-43.

[20] Wauther J. B., Coudert S., Grallien E., Jonard M., Ponnette Q., 2008. Carbone stock in rubber trees plantation in Western Ghana and Mato Grosso (Brazil). For. Ecol. Manag, 255: 2347-2361.

[21] Dinno A., 2015. Nonparametric pairwise multiple comparisons in independent groups using Dunn's test. The Stata Journal, 15 (1), pp. 292 – 300.

[22] Munasinghe E. S., Rodrigo V. H. L., Karunathilake P. K. W, 2011. Carbon sequestration in mature rubber (*Hevea brasiliensis* Muell. Arg.) plantations with genotypic comparison. J. Rubber Res., 91: 36-48.

[23] Juliantika Kusdiana A. P, Alamsyah A., Hanifarianty S., Wijaya T., 2015. Estimation CO_2 Fixation by Rubber Plantation. *In*: 2nd International Conference on Agriculture, Environment and Biological Sciences (ICAEBS'15). Bali, 16-17 August 2015, pp. 16-18.

[24] Kongsager R., Napier J., Mertz O., 2013. The carbon sequestration potential of tree crop plantations. Mitig. Adapt. Strateg. Glob. Change, 18:1197-1213.

[25] Adingra O. M. M. A., Kassi N. J., 2016. Dynamique de la végétation de Bamo et stocks de carbone dans la mosaïque de végétation. European Scientific Journal, 12 (18): 359 – 374

Chapter 6

Elucidation of Some Ecological Traits of Carabids (Coleoptera: Carabidae) Inhabiting Kakuma Campus Grassland, Kanazawa City, Japan

Shahenda Abu ElEla Ali Abu ElEla,
Wael Mahmoud ElSayed and Nakamura Koji

Abstract

Although adult feeding habits and food requirements are currently and reasonably well known for many coleopteran species, still some carabid species are with peculiar feeding guilds. Although many studies have shown a relationship between morphology of mandibles and feeding behavior in different taxal group, still many aspects concerning the feeding behavior of carabids are promising. An assemblage of carabid species was collected from Kakuma Campus grassland in Kanazawa City, Japan. These species were represented by five subfamilies and nine tribes where the highest number of tribes (3 tribes) was confined to subfamily Harpalinae. The collected carabid assemblage was subjected to mandibular analysis and being categorized into two main groups; carnivorous and omnivorous species. Homologies among mandibular characteristics and functional adaptations of the mandible were also proposed to explore how the interaction network of carabids can affect their behavior in different habitats.

Keywords: Cleoptera, Carabidae, manidbles, morpho-ecological, feeding guild, carnivourous, omnivorous

1. Introduction

Coleoptera possess relatively well-known taxonomy and ecological functions, specialized habitat requirements and considered as one of the most diverse groups of insects [1–3]. It was declared that Carabidae is a megadiverse species in coleopteran family with around 33,920 valid species world-wide [3]. They are one of the dominant aboveground invertebrates in diverse pastures and natural grasslands, and possess high abundance and species diversity at soil surface [4–6], thus they are functionally important group [5, 6].

Ground beetles are known for their long legs and powerful mandibles which enable them to be voracious predators, important for the biological control of insect pests on farms [6, 7]. The adult beetles hunt primarily on the soil surface, but will occasionally climb into the foliage in search of food [7].

While many of these species' diets are known, particularly among the Carabinae, the majority of species showed unknown feeding preference [6]. Attempts using morphological characteristics of the mandibles to investigate the feeding preference of different species of carabids have been made [6–12].

The subject of what carabids eat is not new, and numerous researchers have investigated the feeding habits and preferences of a diverse number of carabids [6, 12]. Their studies have offered information on the feeding habits and carabids' feeding preferences [12].

The current study tries to find an answer for the original question: do mandibular morphological characters of adult carabid beetles could give a prediction on their feeding preferences?

Based upon feeding observations by Forbes, as probably the first to examine and describe the mandibles of carabids, who surmised that carabids were herbivorous as well as carnivorous [13]. Other researchers have supported these observations and well establishing carnivory, granivory and herbivory feeding behavior by carabids [14–22]. However, many carabid species may have seasonal diets, being carnivorous during part of the year and largely granivorous or herbivorous at other times [20].

In general, carabid mandibles are mostly similar [23], they were described as "three sided pyramids" and Jeannel [23] was one of the first who developed the nomenclature for the teeth and ridges (**Figure 1**). Indeed, the terminology was then reviewed and expanded by Acorn and Ball [10]. They described the array of teeth and elevations observed on the mesial margins of the mandibles as a series of parallel ridges separated by occlusal grooves. The terebral ridges posterior to the incisors shear; the retinacular teeth and ridges, may also shear, or, act as a compacter [10]. The variable basal region may have one or more teeth or ridges for additional reduction of food, or have a flattened basal face. A basal face may or may not support a basal brush. The basal brush of hairs may be extensive, or, confined to a few hairs, that help transfer the food to enter the mouth through the pharynx. Generally in the current study, based on personal observations, the left mandible was comparatively the dominant mandible being longer and sliding over the dorsal side of the right mandible, although, within the species, there were some observed exceptions (**Figure 2**).

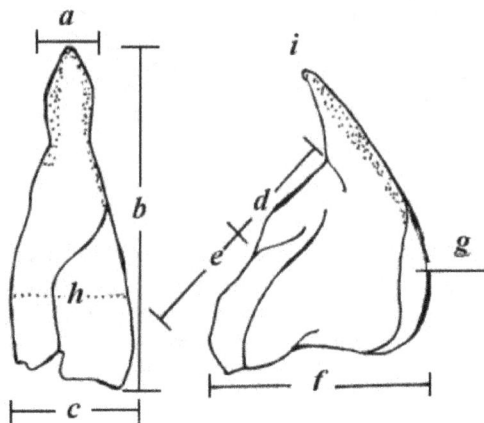

Figure 1.
Diagram of the right mandible of a carabid beetle model showing the main parts: a: Incisor width, b: Mandibular length (l), c: Mandibular base width (w), d: Incisor area, e: Molar area, f: Mandibular width, g: Abductor muscle attachment to the mandible, h: Molar width showing grooves inside the mandible, i: Incisor.

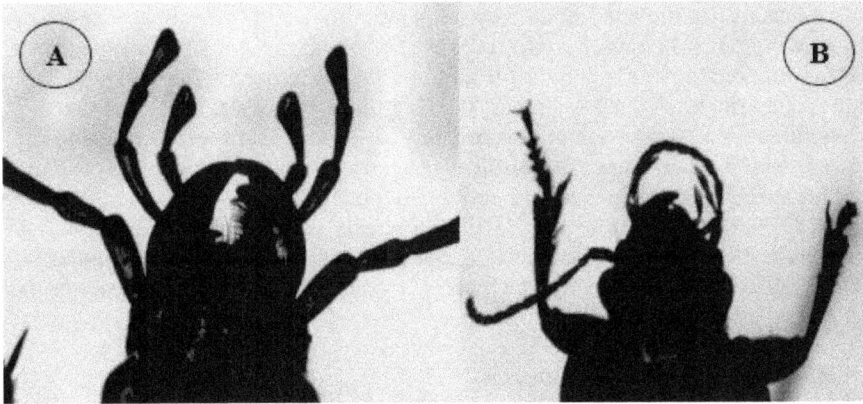

Figure 2.
Difference in mandibular characteristics between predatory carabid (A) represented by Synuchus Synuchus crocatus (bates) showing large and sharp pointed blade-like incisor area compared to an omnivorous carabid (B) represented by Anisodactylus punctatipennis (Morawitz) which showed short, blunt and partially concealed mandibles.

Based on collaborative work, we investigated the ecological traits and specifically the feeding guilds of carabid beetles (Coleoptera: Carabidae) in the grasslands ecosystem located in Kakuma Campus within Kanazawa University, Kanazawa City, Ishikawa prefecture, Japan.

2. Materials and methods

2.1 Study area

The survey of carabid assemblage was conducted in Kakuma Campus grassland - 36.546 N & 136.708 E - within Satoyama area of Kanazawa City, Ishikawa Prefecture, Japan. Kanazawa city is located on the area facing Japan Sea being boarded by the Japan Alps, Hakusan National park and Noto Peninsula National Park. Kanazawa city sits between two rivers - Sai and Asano Rivers - covering an area of ca. 467.77 km². Satoyama, as a part of Kanazawa; covers an area of ca. 74 ha and is located at 150 m altitude, 5 km southeast from the city center. The area comprises various habitat types ranging from secondary forests dominated by *Quercus serrata*, *Quercus variabilis*, *Phyllostachys pubescens*, and *Cryptomeria japonica*.

2.2 Sampling protocol

Specimens were collected primarily from unbaited pitfall traps. At the sampling site, 15 unbaited pitfall traps were installed as trapping tools and spaced about 1 m apart along a transect running north to south through the center of each survey site. The total number of traps in all sampling habitats was 75. In most excursions, sampling of carabids was performed during the days in the middle of the month especially sunny days. Traps were installed in the soil to cover the period from early May till late November, 2019.

The traps consisted of white polyethylene beakers (13.5 cm deep, Ø 9 cm). These beakers were primed with 10% ethylene glycol and we added few drops of ordinary detergent to reduce surface tension. Three wooden sticks were drilled around each pitfall trap (11 cm below the upper brim) and a plastic beaker cover was mounted above each trap to prevent flooding by frequent rainfall and to minimize the damage that could be caused to the traps by the falling leaves or small twigs. The disturbance caused by placing the pitfall traps was minimized and the vegetation around the traps was kept intact. The 'digging in' effect was thus considered negligible and the traps were set immediately [6, 24, 25]. Traps were left open for two consecutive days and in the third day; each trap was emptied from its content and the specimens caught were preserved in Renner's solution (40% ethanol, 30% water, 20% glycerin, 10% acetic acid) [26]. Preserved specimens were then brought back to the laboratory for identification, counting and sorting. To reduce the variability caused by sampling error, only one of the authors (W.M.E.) was responsible for making counts in this study.

2.3 Identification and nomenclature

Carabids were identified to species level and the used nomenclature was in accordance with the key given by Nakane [27]. In addition, the collected species were compared with already identified museum specimens in Kanazawa University for further confirmation. Collected specimens of carabids were deposited and cataloged in Kanazawa University repository room. These specimens were kept in special boxes containing small sachets enclosing naphthalene-coated tablets for further specimen protection against moths and other destructive pests. These sachets were checked regularly and renewed whenever needed.

2.4 Abundance code

Carabid species were categorized into three abundance codes according to the cumulative number of collected specimens during the study period. These abundance code are: Rare ≤5, O: 5 < Occasional ≤15 and A: Abundant >16 individuals.

2.5 Body size

Morphometric measurements of collected carabid species using Vernier® caliper micrometer (precision ±0.10 cm) were performed. Measurements were performed from the tip of the labrum to the extremity of the pygidium and carabids were classified into three body size groups: small (≤ 5 mm), medium (5 mm < body length < 15 mm) and large species (≥15 mm).

2.6 Mandibular analysis and characteristics

All specimens were dissected retaining the head capsule possessing the attaching mandibles. The head capsules were mounted on double-sided tape on slides trays. The head capsules were then precisely dissected retaining the mandibles for further analysis and to take images. Mandibles were lightly brushed with 80% ethanol then by distilled water in an effort to remove most of the soil particles and debris adhered to the mandibles. After air-drying, specimens were examined under Stereo-fluorescence microscope (Nikon® SMZ800 series) equipped with digital camera and TFT LCD Nikon® monitor where illumination was provided from double gooseneck illuminator (Olympus® HLL-301). Syncroscopy Auto-Montage system was taken for photography (Kanazawa University, Laboratory of Biodiversity). All mandibular images were saved and stored as jpeg files for morphological measurements using Image J 1.45 software.

The mandibles were identified morphologically as left and right mandibles when viewed dorsally with the head forwarded to the top of the examining slide.

The main measurements were taken of both mandibles: length (l) and width (w) and compared as ratios (l/w). Measurements were performed in millimeter (mm). Mandibular length was considered from the exterior edge of the dorsal mandibular condyle to the farthest point of the mandibular incisor. On the other hand, the mandibular width was measured from the exterior edge of the dorsal mandibular condyle to the point where attachment for the abductor muscle could be observed (**Figure 1**).

2.7 Feeding guild

From the structure and morphological adaptations of the mandibles, two guilds were mainly assigned: predators (sharp incisors and long terebral ridge) and omnivorous species (dull incisors with short terebral ridge) [6]. Our analyses of mandibular morphology were compared with previous literature whenever data are available.

The feeding guild was predicted from mandibular morphology (**Figure 1**) and compared with previous reports whenever information were available.

3. Results

3.1 Carabid assemblage

A total of 120 individuals of different carabid species belonging to 17 species were recorded from Kakuma Campus grassland. These individuals belonged to five subfamilies and eight tribes (**Table 1**). Subfamily Harpalinae proved to possess the highest number of tribes (3 tribes: Anisodactylini, Harpalini and Zabrini) compared to other observed subfamily in the study area. On the other hand, two subfamilies (Carabinae and Zabrinae) harbored only a single tribe each (Carabini and Callistini, respectively). Subfamily Pterostichinae proved to harbor the highest number of individuals (69 carabid individuals) and this was followed by subfamily Zabrinae (33 individuals). The least number of carabids (2 individuals) was confined to Subfamily Bembidiinae. The highest number of species in one tribe was observed in Callistini (4 species) (**Table 1**).

In general, Kakuma Campus grassland revealed that the study site relatively possessed poor carabid assemblage. According to the cumulative number of individuals of each carabid species; the assemblage showed that rare species were the dominant code where this code comprised ca. 47.1% of the assemblage (**Table 1**). On contrary, the Abundant code (A) comprised ca. 11.76% of carabids co-occurring in Kakuma Campus grassland (**Table 1**).

3.2 Body size

There was a large difference between the number of carabid species with medium-sized bodies and those of other sizes. The majority of carabid species had a medium-sized body. There were 9 medium-size carabids out of 17 species, representing ca. 53% of the total observed species. Large and small-sized species were rare (only 3 species, representing ca. 17.6% of the total observed species) as indicated in **Table 1**. Thus, generally on the habitat level, grassland in Kakuma Campus showed a predominance of species with a medium body size (**Table 1**). On the other hand, on the subfamily level, medium-size carabids were distributed among three subfamilies (Harpalinae, Pterostichinae and Zabrinae) as indicated in **Table 1**. It was observed that small-size carabids were rare in term of number of individuals and were distributed in only two subfamilies (Subfamily Bembidiinae and Harpalinae) and it was apparent that the

Taxa	Individual	Ecological traits*		
		I	II	III
Subfamily Bembidiinae				
Tribe Bembidiini				
Bembidion koikei (Habu et Baba)	1	R	S	Omn
Tribe Tachyini				
Tachyura nana (Gyllenhal)	1	R	S	Omn
Subfamily Harpalinae				
Tribe Anisodactylini				
Anisodactylus punctatipennis (Morawitz)	3	R	M	Omn
Anisodactylus sadoensis (Schauberger)	10	O	M	Omn
Tribe Harpalini				
Harpalus sinicus (Hope)	3	R	M	Omn
Tribe Zabrini				
Amara congrua (Morawitz)	1	R	S	Omn
Subfamily Carabinae				
Tribe Carabini				
Carabus dehaanii punctatostriatus (Bates)	6	O	L	Car
Leptocarabus procerulus (Chaudoir)	3	R	L	Car
Subfamily Pterostichinae				
Tribe Platynini				
Synuchus Synuchus crocatus (Bates)	19	A	M	Car
Synuchus Synuchus cycloderus (Bates)	13	O	M	Car
Synuchus Synuchus dulcigradus (Bates)	11	O	M	Car
Tribe Pterostichini				
Pterostichus polygenus (Bates)	16	A	L	Car
Pterostichus yoritomus (Bates)	10	O	M	Car
Subfamily Zabrinae				
Tribe Callistini				
Chlaenius costiger (Chaudoir)	5	R	L	Car
Chlaenius ocreatus (Bates)	7	O	M	Car
Chlaenius pallipes (Gebler)	6	O	M	Car
Haplochlaenius costiger (Chaudoir)	5	R	L	Car

*Ecological trait:
I- *Abundance code (R – rare, O – occasional, F – Frequent, A – abundant).*
[R: Rare ≤5, O: 5 < Occasional ≤15, A: Abundant >16].
II- *Body size: S: Small, M: Medium and L: large (see text for more details).*
III- *Feeding category: Car: Carnivorous, Omn: Omnivorous.*

Table 1.
List of carabid species inhabiting different grasslands of Kakuma campus with their subfamily, tribe, abundance and ecological trait.

small-size carabids were singleton species (**Table 1**). On the other hand, the number of large-size carabids fell in the middle of this continuum with five large-size species could be observed and distributed in three subfamilies (**Table 1**) representing almost the third of the total catch (29.4%, **Table 1**).

3.3 Feeding guild

Out of the 17 recorded sampled species; 11 species (64.7%) were carnivorous species while only 6 species (35.3%) were omnivorous (**Table 1**). Most of the carnivorous were belonging to three subfamilies (Carabinae, Pterostichinae, and Zabrinae) with the maximum number of species (5 species) belong to subfamily Pterostichinae. On the other hand, omnivorous species were confined to only two subfamilies (Bembiidinae and Harpalinae) with the highest number of omnivorous species were recorded in subfamily Harpalinae (**Table 1**). In general, 19 individuals of carabid were observed to be omnivorous, whereas the carnivorous feeding habit was possessed by 101 carabid individuals.

Typical carnivorous species were characterized by forward-projecting mandibles, sharp incisors used to pierce and capture prey and a long terebral ridge used to kill and slice prey into pieces (**Figure 2**). Omnivorous species, on the other hand, had a wide molar region for crushing seeds but incisors were blunt and the terebral ridge was short as the shown example, *Anisodactylus punctatipennis* (Morawitz), in **Figure 2**. Thus, omnivorous species have features that are advantageous for seed feeding but reduce the efficiency of feeding on prey.

4. Discussion

The Carabidae is considered one of the six largest families in the order Coleoptera and largest family in the suborder Adephaga, with ca. 33,920 valid species world-wide [3]. Some studies estimated that there are 30,000 specie [28] while other studies estimated that the number of carabid species may reach 40,000 species in the world [29]. Carabids have been extensively collected and studied because they exist in a wide range of habitats and can be relatively abundant, and are often agriculturally pertinent [1, 6]. In the present research, however, Kakuma Campus grassland showed a relatively poor assemblage of carabid species. Similar studies, declared that in one-year studies 20–35 species are found in the qualitative structure of ground beetles of cultivation fields in Central Europe Basedow et al. (1976) and Thiele (1977) [30, 31]. Moreover, in the Subcarpathian region, 54 Carabidae species were recorded with the average number of carabids species per site was 15 species [32], and the review publication reported a relatively low number of carabid species in which 12 12 species were recorded in one site out of 21 identified species in total studied sites [32].

Kakuma Campus grassland, as a part of Satoyama, was subjected to relatively low anthropogenic disturbances over a considerable time. These disturbances were focused mainly on regular monthly mowing. Prior excursions in the selected site revealed that the area is relatively with poor carabid diversity parameters compared to other sites within Satoyama, for example Kitadan area within Satoyama landscape [6]. However, more data are required to clarify more aspects concerning carabid assemblage (some of the data on other sites concerning this project were not published yet for comparisons).

Diverse studies showed that there is a long history of success in using carabids to signal environmental change [4, 8, 33–35]. Moreover, changes in landscape such as fragmentation [36, 37], recreational use [38], urbanization [39, 40], forest management [35, 41].

We assume that poor carabid assemblage as being represented by dominant rare species and relatively scarcity of abundant species in Kakuma Campus grassland as a result of regular mowing which led to relatively poor assemblage [42]. This view could be supported by other related studies which suggested that carabids could been used as indicators of large-scale environmental changes [43], and predictors of future landscape changes [4, 44].

Moreover, worth noting that fragmentation of continuous habitats, as the case of Satoyama, into many small patches or relatively small habitats as a result of anthropogenic impact such as urbanization and/or cultivation may affect the co-occurring carabids making some populations highly isolated [4, 45, 46].

In the present study, carabid species possessed diverse morphological traits which were focused mainly on body size and mandibular characteristics. There were numerous investigations dealing with morphological traits of carabids and their life strategies among different habitats [47–57].

The co-occurring carabids were predominated by small and medium-size species and large-size species were the least dominant. Other studies support our findings in which they suggested that large-bodied carabids were missing from many small islands and generally less abundant when present, with the opposite true for smaller-bodied species (Bell et al. 2017). Similar findings have been reported for carabids in relation to size of forest patches [10, 44].

Nonetheless, some studies stated that the prevailing tendency towards a relatively higher number of both small~medium size carabids is a typical phenomenon, characteristic of habitats that are subjected to external factors [58, 59].

Feeding guild was an additional important trait used to analyze the structure of carabidofauna [6, 32]. The collected carabids were dominated by carnivorous species. A similar study revealed that zoophages (predators or carnivours in other terms) were predominant in the entire assemblage [32].

Diverse studies ranked carabids into three main categories according to their pattern of food intake: oligophagous predators, polyphagous predators and phytophages. Granivory habit was, consequently, confirmed by a wide-range of diverse studies [8–10, 16–18, 22, 29, 30, 60–66]. Other studies stated that enormous number of carabid individuals may exist in farm fields, in communities of carnivore and granivores and, more in deed, obligate omnivore guilds [67].

Typical carnivorous species were characterized by forward-projecting mandibles, sharp incisors used to capture and pierce the prey. The mandible was with a long terebral ridge used to kill and slice prey into pieces. Diverse studies showed that the diet of carabids included Collembola, earthworms, nematodes, slugs, snails, aphids, eggs and larvae of Diptera and Coleoptera, Lepidoptera pupae and seeds of herbaceous plants [2, 20, 68, 69]. Hence, carabids are crucial predators in agricultural landscapes feeding on a wide-range of preys [70].

Beetles use their mandibles for prey capture and the forces created by the mandible tips are used to hold prey and pierce the integument [71, 72].

Omnivorous species, on the other hand, had a wide molar region for crushing seeds but incisors were blunt and the terebral ridge was short [6]. Thus, omnivorous species have features that are advantageous for seed feeding but reduce the efficiency of feeding on prey [6, 12].

Although adult feeding habits and food requirements are currently and reasonably well known for many carabid species, still some carabid species with peculiar feeding guilds. Some morpho-functional studies have shown a relationship between morphology and feeding behavior in the larval stage [73]. Some morpho-ecological types were defined in the European temperate zone and places most Harpaline and Zabrine, especially the larval stage, with a phytophagous diet pattern into the morpho-ecological types [73]. Nevertheless, these types were minimized into two simpler main categories: a – spermophagous (seed predators); b – c-shaped harpalines, excluding the subtribe Ditomina [74].

Ultimately, the apparent similarities between these mandibles and the jaws of various mammals are remarkable to consider [10, 31]. The incisor area of the mandibles of beetles is related to the cutting incisors encountered in rodents and lagomorphs. The posterior molars of most mammals, on the other hand, are geared to adapt the function of grinding function requiring cusps occluding into basins. This does not appear to be the case in the carabids reviewed here (based on personal observations).

The study of mandibular traits in carabids is obvious to be of significant interest, since they may be beneficial in systematic research and can be linked to feeding patterns employing simple functional explanations [6].

The gathered data on the comparatively small sample of surveyed carabid taxa give only a hint concerning the whole story of evolution of mandibular morpho-functional characteristics in adult carabids.

We hope that other studies will find the morpho-functional studies of carabid mandibles.

It is hoped that more researchers will find the study of mandibles is rewarding and will contribute to the advancement of our knowledge. The study offered here opens the door for more studies to analyze more mandibles from more carabid taxa. We believe that carabidologists would benefit greatly from these studies in their efforts to understand the evolution and adaptations of carabid beetle taxa.

5. Conclusion

In conclusion, further studies would benefit from the examination of additional carabid taxa since making a general connection between mandibular morphological dentition and dietary pattern - generally feeding guild- is far from precise evaluation. Other evolutionary relative lineages among Carabidae are required in order to better address their precise feeding guilds. From that view, incorporation of mandibular morpho-ecological features studies together with phylogenetic analysis are recommended. Consequently, further examination of the gut contents of carabid taxa in conjunction with laboratory investigations and precise observations of feeding behavior in diverse habitats could be employed as confirmation cues for not placing carabids in an ambiguous feeding guild. To summarize, studies merely on morphological characteristics of carabid mandibles are difficult to interpret without an understanding of the functional consequences of variations in mandibular configurations in different carabids habitats. This would reveal some hidden aspects that could not be deduced from the morphological characters of the mandibles if they were adopted alone.

Acknowledgements

The authors wish to acknowledge the staff at the Laboratory of Ecology, Graduate School of Natural Science and Technology, Kanazawa University for their sincere assistance and prompt provided facilities. Also, sincere and warm gratitude is extended to Prof. Dr. Nakamura Koji (Kanazawa University, Japan) for his keen and endless hospitality and encourages for completing the research and writing the manuscript.

Conflict of interest

The authors declare that there are no conflict of interest associated with this article.

Author details

Shahenda Abu ElEla Ali Abu ElEla[1]*, Wael Mahmoud ElSayed[1] and Nakamura Koji[2]

1 Department of Entomology, Faculty of Science, Cairo University, Egypt

2 Graduate School of Natural Science and Technology, Kanazawa University, Kanazawa, Japan

*Address all correspondence to: shosho_ali76@yahoo.com

IntechOpen

References

[1] Lövei GL, Sunderland KD. Ecology and behavior of ground beetles (Coleoptera: Carabidae). Annual Review of Entomology. 1996;**41**:231-256

[2] Kromp B. Carabid beetles in sustainable agriculture: A review on pest control efficacy, cultivation impacts and enhancement. In: Paoletti MG, editor. Invertebrate Biodiversity as Bioindicators of Sustainable Landscapes. Amestrdam, Netherlands: Elsevier Science; 1999. pp. 187-228

[3] Bousquet Y. Catalogue of geadephaga (Coleoptera, Adephaga) of America, North of Mexico. ZooKeys. 2012;**245**: 1-1722

[4] Niemelä J, Kotze J, Ashworth A, Brandmayr P, Desender K, New T, et al. The search for common anthropogenic impacts on biodiversity: A global network. Journal of Insect Conservation. 2000;**4**:3-9

[5] Niemelä J, Koivula AM, Kotze ADJ. The effects of forestry on carabid beetles (Coleoptera: Carabidae) in boreal forests. Journal of Insect Conservation. 2007;**11**:5-18

[6] Mahmoud ESW. Diversity and Structure of Ground Beetle (Coleoptera: Carabidae) Assemblages in Satoyama. Kanazawa University Library; 2010. p. 152

[7] Snyder WE. Give predators a complement: Conserving natural enemy biodiversity to improve biocontrol. Biological Control. 2019;**135**:73-82

[8] Lindroth CH. Die Fenoskandischen Carabidae III Allgemeiner Teil. Kgl. Sweden: Vetenskaps Vitterhetssamhaellets Handl F6, Ser. B 4; 1949. p. 911

[9] Forsythe TG. Feeding mechanisms of certain ground beetles (Coleoptera: Carabidae). Coleopterists Bulletin. 1982;**36**(1):26-73

[10] Acorn JH, Ball GE. The mandibles of some adult ground beetles: Structure, function, and the evolution of herbivory (Coleoptera: Carabidae). Canadian Journal of Zoology. 1991;**69**:638-650

[11] Ball G, Shpeley D, Acorn J. Mandibles and labrum-epipharynx of tiger beetles: Basic structure and evolution (Coleoptera, Carabidae, Cicindelitae). ZooKeys. 2011;**147**:39-83

[12] Ingerson-Mahar M. Relating diet and morphology of the head, mandibles and proventriculus in adult carabid beetles [thesis]. New Jersey: Graduate School, New Brunswick Electronic Theses and Dissertations, Rutgers, The State University of New Jersey; 2014. p. 105

[13] Forbes SA. The food relations of the Carabidae and Coccinellidae. Illinois State Laboratory of Natural History, Bulletin. 1883;**1**:33-64

[14] Webster FM. *Harpalus caliginosus* as a strawberry pest, with notes on other phytophagous Carabidae. Canadian Entomologist. 1900;**32**:265-271

[15] Shough WW. The feeding of ground beetles. American Midland Naturalist. 1940;**24**:336-344

[16] Davies MJ. The contents of the crops of some British carabid beetles. Entomologists Monthly Magazine. 1953;**95**:25-28

[17] Skuhravy V. Diet of field carabids. Casopis Ceskoslov Spolecnosti Entomologick'e. 1959;**56**(1):1-19

[18] Johnson NE, Cameron RS. Phytophagous ground beetles. Annals of the Entomological Society of America. 1969;**62**(4):909-914

[19] Lund RD, Turpin FT. Carabid damage to weed seeds found in Indiana cornfields. Environmental Entomology. 1977;**6**(5):695-698

[20] Tooley, J. and G. E. Brust, 2002. Weed seed predation in carabid beetles, in, Agroecology of Carabid Beetles, J. Holland, ed., Intercept Limited, Andover, UK, pgs 215-229

[21] Honek A, Martinkova Z, Saska P. Post-dispersal predation of Taraxacum officinale (dandelion) seed. Journal of Ecology. 2005;**93**:345-352

[22] Honek A, Martinkova Z, Saska P, Pekar S. Size and taxonomic constraints determine the seed preferences of Carabidae. Basic and Applied Ecology. 2007;**8**:343-353

[23] Jeannel R. Monographie des Trechinae Morphologie comparée et distribution geogrphique d'un groupe de Coleoptères. Abeille. 1926;**32**:285-293

[24] Greenslade P. Sampling ants with pitfall traps: Digging-in effects. Insectes Sociaux. 1973;**20**:343-353

[25] Botes A, Mcgeoch MA, Chown L. Ground-dwelling beetle assemblages in the northern Cape Floristic Region: Patterns, correlates and complications. Austral Ecology. 2007;**32**:210-224

[26] Renner K. Coleopterenfänge mit Bodenfallen am Sandstrand der Ostseeküste—ein Beitrag zum Problem der Lockwirkung von Konservierungsmitteln. Faunistisch-Ökolog. Mitt. 1982;**5**:137-146

[27] Nakane T. Carabidae. In: Nakane T, editor. Iconographia Insectorum Japonicorum Color Naturali edita. Vol. II (Coleoptera). Hokuryūkan, Shōwa, Tokyo, Japan; 1978 (In Japanese)

[28] Lawrence JF. Coleoptera. In: Parker SP, editor. Synopsis and Classification of Living Organisms.

Vol. 2. New York: McGraw-Hill, 1232 pp; 1982. pp. 482-553

[29] Larochelle A. The food of carabid beetles (Coleoptera: Carabidae, including Cicindelinae). Fabreries, Supplement 5. Canada: Association des Entomologistes Amateurs du Québec; 1990

[30] Thiele HU. Carabid Beetles and their Environments. Berlin: Springer; 1977. p. 369

[31] Basedow T, Borg Å, de Clercq R, Nijveldt W, Scherney F. Untersuchungen über das Vorkommen der Laufkäfer (Col.: Carabidae) auf Europäischen Getreidefeldern (Studies on the occurence of Carabidae in European wheat fields). Entomophaga. 1976;**21**(1):59-72 (In German, abstract in English)

[32] Czerniakowski ZW, Olbrycht T, Konieczna K. Ground Beetles (Coleoptera: Carabidae) found in conventional potato cultivations (Solanum tuberosum L.) in the subcarpathian region. Applied Ecology and Environmental Research. 2020;**18**(2):2109-2128

[33] Stork N, editor. The Role of Ground Beetles in Ecological and Environmental Studies. Andover, Hampshire, UK: Intercept Publications; 1990. pp. 424

[34] Desender K, Dufrêne M, Loreau M, Luff ML, Maelfait J-P, editors. Carabid Beetles: Ecology and Evolution. Dordrecht, Netherlands: Kluwer Academic Publishers; 1994. p. 474

[35] Niemelä J. From systematics to conservation – carabidologist do it all. Annales Zoologici Fennici. 1996;**33**:1-4

[36] Den Boer PJ. Dispersal Power and Survival: Carabids in a Cultivated Countryside. Vol. 14. Netherlands: Landbouwhogeschool Wageningen Miscellaneous Papers; 1977. pp. 1-190

[37] Brandmayr P. Insect communities as indicators of anthropogenic

modifications of the landscape and for territory planning: Some results by carabid beetles. Atti XII Congr. Naz. Ital. Ent. Roma. 1980;**1**:263-283

[38] Emetz VM. Mnogoletnyaya dinamika pokazatelei izmenchivosti gruppirovok imago zhuzhelitsy *Pterostichus oblongopunctatus* F. (Coleoptera, Carabidae) polimorfnomu priznaku (chislo yamok na nadkryl'yakh) na rekreatsionnom i maloposeshchaemom uchastkakh dubrav. Entomologicheskoe obozrenie. 1985;**64**:85-88 [in Russian]

[39] Czechowski W. Occurrence of carabids (Coleoptera, Carabidae) in the urban greenery of Warsaw according to the land utilization and cultivation. Memorabilia Zool. 1982;**39**:3-108

[40] Klausnitzer B. Faunistisch-¨okologische Untersuchungen ¨uber die Laufk¨afer (Col., Carabidae) des Stadtgebietes von Leipzig. Entomologische Nachrichten und Berichte. 1983;**27**:241-261

[41] Spence JR, Langor DW, Niemelä J, C'arcamo, H. A. and Currie, C. R. Northern forestry and carabids: The case for concern about old-growth species. Annales Zoologici Fennici. 1996;**33**:173-184

[42] Schwerk A, Dymitryszyn I. Mowing intensity influences degree of changes in carabid beetle assemblages. Applied Ecology and Environmental Research. 2017;**15**(4):427-440

[43] Penev L. Large-scale variation in carabid assemblages, with special references to the local fauna concept. Annales Zoologici Fennici. 1996;**33**: 49-64

[44] Müller-Moetzfeld G. Laufkäfer (Coleoptera, Carabidae) als pedobiologische Indikatoren. Pedobiologia. 1989;**33**:145-153

[45] McGeoch MA. The selection, testing and application of terrestrial insects as bioindicators. Biological Reviews. 1998;**73**:181-201

[46] Jelaska LS, Durbesic P. Comparison of the body size and wing form of carabid species (Coleoptera: Carabidae) between isolated and continuous forest habitats. Annales de la Société Entomologique de France. 2009;**45**(3):327-338

[47] Šustek Z. Changes in body size structure of carabid communities (Coleoptera, Carabidae) along an urbanization gradient. Biológia (Bratislava). 1987;**24**:145-156

[48] Gutiérrez D, Menéndez R. Patterns of the distribution, abundance and body size of carabid beetles (Coleoptera: Caraboidea) in relation to dispersal ability. Journal of Biogeography. 1997;**24**:903-914

[49] Szyszko J, Vermuelen HJW, Klimaszewski M, Schwerk A. Mean Individual Biomass (MIB) of ground beetles (Carabidae) as an indicator of the state of the environment. In: Brandmayr P, Lövei G, Brandmayr TZ, Casale A, Vigna TA, editors. Natural history and applied ecology of carabid beetles, Pensoft publishers. Moscow: Sofi a; 2000. pp. 289-294

[50] Ribera I, Dolédec S, Downie IS, Foster GN. Effect of land disturbance and stress on species traits of ground beetle assemblages. Ecology. 2001;**82**: 1112-1129

[51] Niemelä J, Kotze DJ, Venn S, Penev L, Stoyanov I, Spence J, et al. Carabid beetles assemblages (Coleoptera, Carabidae) across urban-rural gradients: An international comparison. Landscape Ecology. 2002;**17**:387-401

[52] Braun SD, Jones TH, Perner J. Shifting average body size during regeneration after pollution - a case study using

ground beetle assemblages. Ecological Entomology. 2004;**29**(5):543-554

[53] Weller B, Ganzhorn JU. Carabid beetle community composition, body size, and fluctuating asymmetry along an urban-rural gradient. Basic and Applied Ecology. 2004;**5**:193-201

[54] Magura T, Tóthmérész B, Lövei GL. Body size inequality of carabids along an urbanization gradient. Basic and Applied Ecology. 2006;**7**:472-482

[55] Elek Z, Lövei GL. Patterns in ground beetle (Coleoptera: Carabidae) assemblages along an urbanization gradient in Denmark. Acta Oecologica. 2007;**32**(1):104-111

[56] Šerić Jelaska L, Blanuša M, Durbešić P, Jelaska SD. Heavy metal concentrations in ground beetles, leaf litter, and soil of a forest ecosystem. Ecotoxicology and Environmental Safety. 2007;**66**(1):74-81

[57] Gaublomme E, Hendrickx F, Dhuyvetter H, Desender K. The effects of forest patch size and matrix type on changes in carabid beetle assemblages in an urbanized landscape. Biological Conservation. 2008;**141**:2585-2596

[58] Pałosz T. Intensive technologies in agriculture in relation to the ground beetle fauna. Ochrony Roślin. 1995;**39**(5):8 (in Polish)

[59] Leśniak A. Methods of analysis of ground beetle (Col., Carabidae) assemblages in zooindication of ecological processes. – Evaluation of forest ecosystems by zooindication methods. Warszawa: Wyd SGGW. 1997:29-41

[60] Burmeister F. Biologie, Okologie und Verbreitung der Europaeischen Kaefer. I. Band: Adephaga, Caraboidea. Krefeld, Germany: Goecke Verlag; 1939. p. 206

[61] Dawson N. A comparative study of the ecology of eight species of fenland Carabidae (Coleoptera). Journal of Animal Ecology. 1965;**34**:299-314

[62] Burakowski B. Biology, ecology and distribution of *Amara pseudocommunis* Burak. (Coleoptera, Carabidae). Annales Zoologici. 1967;**24**(9):485-523

[63] Manley GV. A seed-cacheing carabid (Coleoptera). Annals of Entomological Society of America. 1971;**64**:1474-1475

[64] Hurka K. Carabidae of the Czech and Slovak Republics. Zlin, Czech Republic: Kabourek; 1996. p. 565

[65] Jørgensen HB, Toft S. Food preference, diet dependent fecundity and larval development in *Harpalus rufipes* (Coleoptera: Carabidae). Pedobiologia. 1997a;**41**:307-315

[66] Jørgensen HB, Toft S. Role of granivory and insectivory in the life cycle of the carabid beetle *Amara similata*. Ecological Entomology. 1997b;**22**:7-15

[67] Alice C, Petit S, Dechaume-Moncharmont F-X, Bohan D. Can obligatory omnivore carabids be useful for the biocontrol of weeds?. 18. European Carabidologist Meeting. Rennes, France: Institut National de la Recherche Agronomique (INRA). FRA; 2017. p. 92

[68] Holland JM. The Agroecology of Carabid Beetles. Andover, Hampshire, UK: Intercept publication Limited; 2002

[69] Holland JM, Luff ML. The effects of agricultural practices on Carabidae in temperate agroecosystems. Integrated Pest Management Reviews. 2000;**5**:109-129

[70] Dinis A, Pereira J, Benhadi-Marín J, Santos S. Feeding preferences and functional responses of *Calathus granatensis* and *Pterostichus globosus* (Coleoptera: Carabidae) on pupae of *Bactrocera oleae* (Diptera: Tephritidae). Bulletin of Entomological Research. 2016;**106**(6):701-709

[71] Wheater CP, Evans MEG. The mandibular forces and pressures of some predacious coleoptera. Journal of Insect Physiology. 1989;**35**:815-820

[72] Mair J, Port GR. Predation by the carabid beetles *Pterostichus madidus* and *Nebria brevicollis* is affected by size and condition of the prey slug *Deroceras reticulatum*. Agricultural and Forest Entomology. 2002;**3**(2):99-106

[73] Sharova IK. Morpho-ecological types of carabid larvae. Zool Zh. 1960;**39**:691-708 [in Russian]

[74] Zetto Brandmayr T, Giglio A, Marano I, Brandmayr P. Morphofunctional and ecological features in carabid larvae: A contribution to distinguish between affinity and convergence. Proc. XX. International Congress of Entomology, Firenze, 28-31 August 1996. Museo Regionale di Scienze Naturali di Torino, Torino. 1998:449-490

Comparison of the Effectiveness of Different Tags in the Sea Urchin *Paracentrotus lividus* (Lamarck, 1816)

Noelia Tourón, Estefanía Paredes and Damián Costas

Abstract

The marking of sea urchins was implemented with the main objective of being able to individually identify the urchins in the natural environment once released. In addition, it's very useful to monitor individuals in studies of growth, movements, development, population dynamics, etc., that develop in the natural environment. Numerous different marking methodologies have been tested for sea urchins, either by physical marking (external and internal labels) or by using chemical marking methods consisting of the use of fluorochromes, which adhere to the calcified structures of the urchin. In this work, 5 different physical marks were used to mark 400 urchins of the *Paracentrotus lividus* species, which were kept for a month at the ECIMAT facilities in Toralla island. The efficacy of the methods used in each case was analyzed, comparing the survival rate and the tag retention rate of the tagged urchins obtained with each tagging methodology.

Keywords: sea urchin, marking, tag, retention rate, survival

1. Introduction

The impoverishment of the health of coastal ecosystems in general increases due to overfishing, which has generated a rapid decrease in resources with repercussions on the economic sustainability of the global fishing sector, also leading to a decrease in biodiversity and a reduction in the food security [1].

The ecological importance of sea urchins is crucial, since they are the major regulators of the biomass of macroalgae on the seabed, which proliferate uncontrollably in habitats where sea urchin populations are depleted or completely disappeared [2–6], producing an imbalance in coastal ecosystems that affects the capture of other fishery products. The sea urchin is very sensitive to extraction due to the low population densities that it presents to the preference of individuals to inhabit shallow waters [7–9], and other factors such as human contamination or disease [10].

On the other hand, its economic importance has increased significantly in recent decades, due to the increase in demand for this product in the world market; conversely, the global catch of sea urchin from fisheries decreased from 117,000 tonnes in 1998 to 69,202 tonnes in 2014 [1]. This decrease was mainly due to the reduction in catches of the large world producers, such as Chile and the United States [11, 12], to counteract the overexploitation to which sea urchins were being subjected, although these measures were not effective for the recovery of natural populations.

Sea urchins' gonads have been used as food since Roman times. Some species of sea urchins have been exploited commercially as food resources since the seventeenth century, being a highly valued resource in the market, where the highest quality gonads are a "gourmet" product that can reach a value of $ 300/kg. Asian countries are the major producers of sea urchins, with a total volume of 73,000 T per year [13], which in economic terms represents a total of between 200 and 300 million dollars [13].

In Europe, the most commercially important species is *Paracentrotus lividus*, Galicia is the main producing region of this species, with a total of 695 T of the sale in the fish markets during 2019 and an average sale price of € 8.28/kg, reaching peaks of up to € 24/kg of sea urchins [14].

As a consequence of this overfishing and the need to restore depleted natural populations, aquaculture production of the main commercial species has increased greatly in recent years [15–17], carrying out large-scale restocking tasks in different regions affected by the overexploitation of natural sea urchin banks around the world.

In order to monitor the urchins released to the natural environment and thus carry out subsequent studies on the effectiveness of the repopulation performed (survival, population dynamics, temporal evolution, etc.), it is necessary to identify the individuals released through the use of different types of individual brands. The tags used must be as less invasive as possible, so that they do not affect the growth or movement of individuals in the environment, and must present the maximum percentage of survival and retention rate possible for later recovery of the marked specimens.

There are various methods for marking sea urchins that have been studied and tested over the years in different species. Some of these studies are listed below, classified according to the type of marking used.

In general terms, the marking studies carried out with sea urchins were based on the use of five types of marks: 1) external labels or other markers of different shapes and colors inserted in the spines [18–23]; 2) tags anchored by a perforation of the urchin's test [18, 24–29]; 3) passive integrated transponder (PIT) tags [30]; [13, 31] coded wire labels or CWT [31]; 5) marking using fluorochromes, such as tetracycline or calcein [32–39]. All the methods used have advantages and disadvantages. The physical marking techniques of urchins, both external and internal, generally have low survival or retention rates of the mark, especially, when released to the natural environment, and may also affect the growth of marked sea urchins, and are not viable for the marking of small individuals [40].

The tags that performed the best so far in terms of survival and retention rate of the mark are the passive integrated transponder tags (PIT Tags), which consist of a cylinder fitted with a copper antenna and a microchip programmed with a number of identification (they contain millions of unique codes for the identification of the marked specimens), although they are not suitable for marking urchins with a test diameter less than 25 mm, and their effectiveness varies according to the species of sea urchin that is marked, reducing the survival rate of tagged urchins once released to the natural environment [41].

Chemical labeling with fluorescent substances works just as well as other physical labeling methods, also allowing the marking of large numbers of urchins of any size by immersing individuals in fluorochrome baths [42] or polyfluorochrome [42], although they present a significant disadvantage with respect to other marking techniques, being necessary the sacrifice of marked urchins to detect the mark, which makes it unfeasible to study the evolution of juveniles released for restocking purposes of overexploited areas. The objective of this work is to analyze the efficacy of 5 different physical marks for the identification of individuals of the species *P. lividus*.

2. Materials and methods

A. In June 2020, 400 juvenile urchins (*P. lividus*) were received from the coast of Cangas do Morrazo (Pontevedra, Galicia): 42° 16′40 ″ N 8° 47′23 ″ W, with an average of size 20 mm in diameter and an average weight of 5.4 g.

The urchins were distributed in boxes of 50 liters of capacity, at a density of 20 urchins per box, with seawater filtered in an open circuit, continuous aeration, and feeding "ad libitum" with brown macroalgae of the genus *Laminaria sp*. The duration of the experiment was 1 month (started on June 23, 2020, ended on July 24, 2020). Five types of different tags were used for sea urchins:

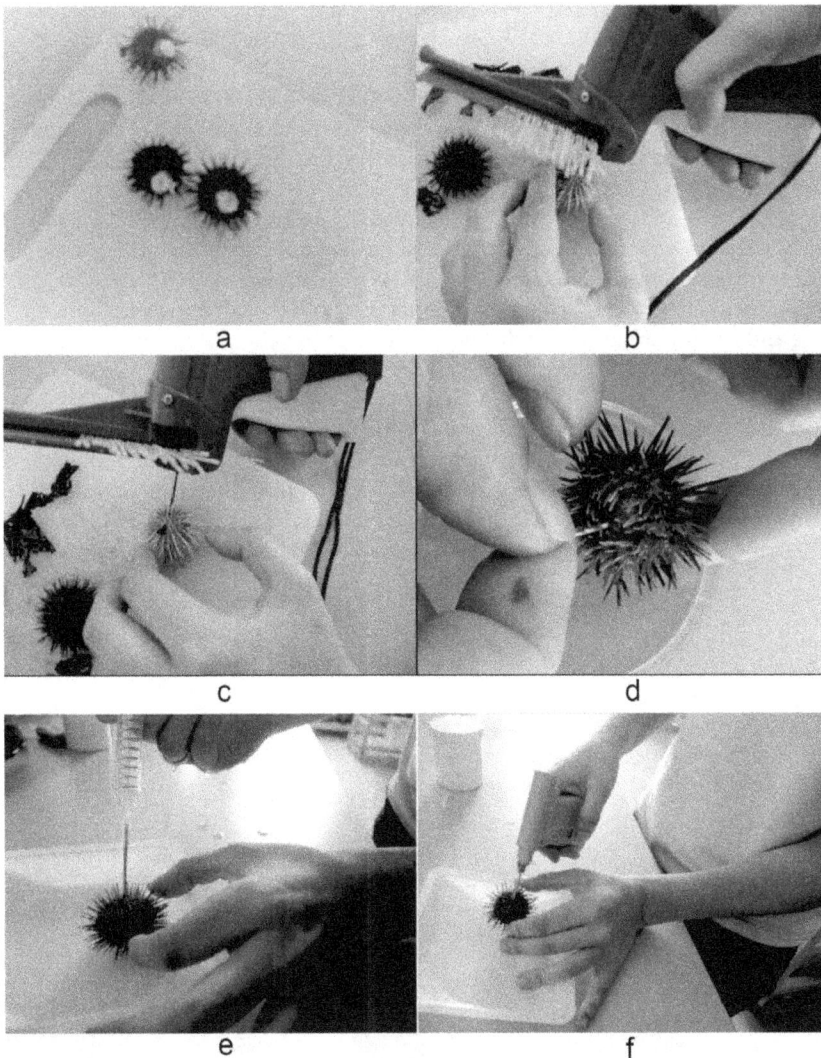

Figure 1.
Different techniques for mechanical marking of sea urchins: a) Hallprint adhesive stickers; b) insertion of T-Bar tags through the peristomial membrane; c) insertion of T-Bar labels by drilling the aboral region of the test; d) insertion of galvanized wire through the peristomial membrane; e) injection of a mini transponder (Trovan brand) through the peristomial membrane; f) injection of PIT Tag (Hallprint brand) through the peristomial membrane.

1. Colored stickers (Hallprint brand): FPN 8x4 (glue on shellfish tag), in 3 different colors (green, purple, and beige) with individual numbering.

2. T-Bar labels (Hallprint brand): TBA (standard anchor T-Bar tag), in three different colors (green, purple, and beige) and individual numbering.

3. Minitransponder (Trovan brand): $1'4 \times 8$ mm, high-performance ISO FDXA glass, with IM-200 1.4 Mini Tradi injector.

4. Pieces of galvanized wire (3–4 mm long and 1 mm thick).

5. PIT Tags (Hallprint brand): FDX Food-safe polymer (2.18 x 11.4 mm).

All the treatments with three replicas per tag and a consistent control of unmarked urchins.

The stickers were adhered to the urchin's test with the help of Loctite glue (**Figure 1**), in an area of the test where the spines of that area were previously sectioned, and the area was dried with absorbent paper. The galvanized wire sections were introduced into the coelomic cavity through the peristomial membrane of the urchin; Trovan Minitransponders and Hallprint PIT Tags were also introduced through the peristomial membrane of the urchin using a specific injector for each type of tag. The T-Bar labels were introduced in two ways into the urchins, half of the labels were introduced through the peristomial membrane and the other half through a hole drilled in the aboral half of the test with the help of a needle.

B. The 270 surviving urchins from the captive tagging experiment were housed in a tray belonging to the polygon of rafts of the San Xosé de Cangas do Morrazo Fishermen's Association (**Figure 2**), on October 9, 2020, with an average size of 18 mm in diameter and an average weight of 2.98 g, in order to obtain the recapture rate of the marks in the natural environment.

The coordinates of the raft are as follows:
Latitude Lenght.
Vertex A 42° 16′ 31" N 08° 43′ 53" W.
Vertex B 42° 16′ 43" N 08° 43′ 30" W.
Vertex C 42° 16′ 42" N 08° 43′ 15" W.
Vertex D 42° 15′ 49" N 08° 43′ 22" W.
Vertex E 42° 15′ 35" N 08° 43′ 59" W.

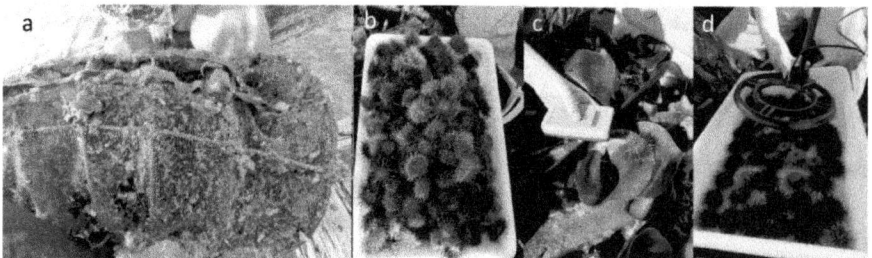

Figure 2.
a) Urchins housed in the raft's lantern; b) urchins taken from the lantern; c) Trovan Minitransponder reader; d) metal detector.

The urchins were fed fortnightly with brown algae of the genus *Laminaria sp.*

The duration of the experiment was 5 months (started on October 9, 2020, ended on April 9, 2021).

3. Results

A. Below, you can see the survival and retention rates of the brand obtained with each type of tag employed.

As can be seen in **Figure 3**, the survival obtained was high with all the tags used, except with the T-Bar labels inserted through a hole drilled in the aboral half of the test, since this perforation did not calcify and produced the death of more than 50% of the marked urchins.

The retention rate obtained with each tag in captive conditions varied greatly, being insignificant in the case of the colored stickers and T-Bar labels introduced through the peristomial membrane of the urchin, which was totally expelled after a few days, and very high in the case of wire sections (93.33%) and Trovan Minitransponders (83.33%), which suggests the adequacy of these two types of marks for marking of individuals of *P. lividus* in captive conditions.

In the case of the Hallprint stickers, it was observed that the urchins detached them with the help of the spines and pedicels, presenting the lowest retention rate of the mark together with the T-Bar labels and the Hallprint PIT Tags, therefore, these marks are not suitable for the identification of urchins released into the wild.

Another drawback observed was that the glue produced abrasion injuries in the area of the urchin's test where it was applied (**Figure 4**), leaving important sequelae to the urchins marked with this technique, although it did not cause the death of the marked individuals.

A Chi-square test (p-value < 0.05) was performed to statistically compare the efficacy of the tags used, resulting in significantly lower survival in urchins labeled

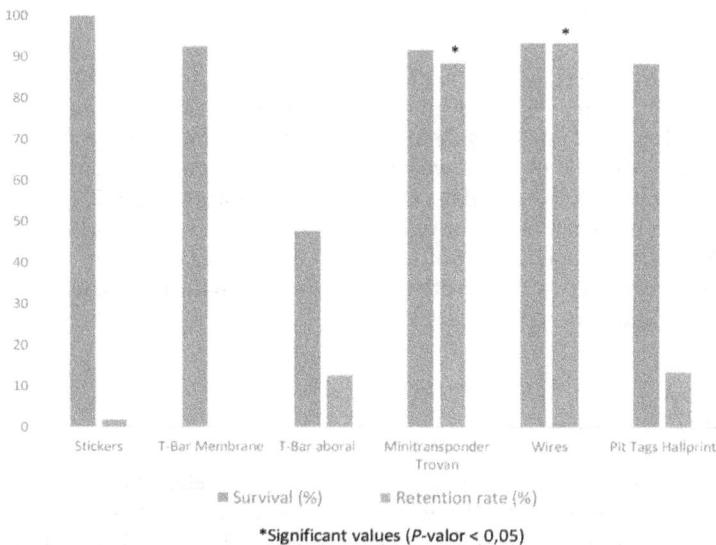

*Significant values (P-valor < 0,05)

Figure 3.
Percentage of survival and retention rate of the urchin's tag with the different tags employed.

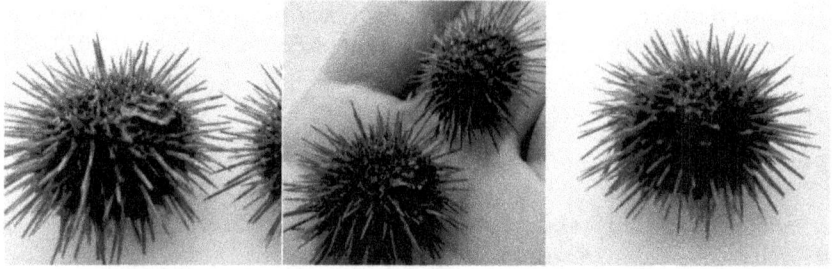

Figure 4.
Abrasive lesions on the urchin's test caused by the glue used to fix the Hallprint stickers.

aborally with T-Bar labels, while the retention rates of the wires and the Trovan Minitransponders were significantly higher than those of the rest of the tags employed.

B. Results obtained with the different labels after housing the marked urchins for 5 months in a culture structure (**Figure 2a**) suspended from a tray belonging to the polygon of rafts of the San Xosé Fishermen's Association, in Cangas do Morrazo (Ría de Vigo).

Recapture rate:

• PIT Tags Trovan: 71%

• PIT Tags Hallprint: 15%

• Stainless steel wires: 0%

• Stickers: 0%

• T-Bar labels: 0%

Total Survival rate: 99,14%

In the second part of the study performed with the urchins housed in a lantern suspended from the raft in the natural environment, very high survival rates were obtained, the total survival rate being 98.89% of the urchins housed in the raft.

The mark-recapture rates obtained were very low after 5 months, except in the case of the Trovan Minitransponders, with which a recapture rate of 71% was obtained, and a mark retention rate of 100% in recaptured urchins, which makes them the most appropriate type of tag for monitoring sea urchins of the *Paracentrotus lividus* species, both in captive conditions and in the natural environment and in short- and long-term studies.

4. Discussion

There are different marking techniques for sea urchins, developed over the last decades, both physical (external and internal) and chemical (different fluoro-chromes). The characteristics that an effective tag must meet for the identification of individuals released to the natural environment are the following: high survival rate of the urchin, high retention rate of the tag for at least a few months, ease of detection on the seabed for divers, ease of identification of tagged individuals, speed of tag application (in order to tag as many sea urchins as possible in a short period of time), and low cost.

Mechanical marking techniques in sea urchins, either by placing external labels of different types or by intraperistomial insertion of internal labels, have the disadvantage of generally presenting low rates of tag retention [18, 43], which may also affect the survival of the marked population. Furthermore, they are not viable to mark small-sized urchins [40] and must be adapted to the morphology of the species to be marked, since there are important differences in the effectiveness of the marks depending on the size of the species, the length of thorns, etc. Tuya et al. [44] used fishing hooks attached to a cork buoy by means of a line to mark long-spined urchins of the *Diadema antillarum* species, obtaining a very high retention rate of the mark between 80 and 90% of marked urchins. Due to the low retention rate generally obtained with external labels, labels inserted in the test began to be used with variable success [25, 26]. Lees [45] tagged *Strongylocentrotus purpuratus* with stainless steel wire and obtained a 92% loss of tagged urchins after 9 months in the wild. Neill [21] marked sea urchins with anchor tags designed for marking fish and provided with a specific numbering, obtaining a very low survival rate from 11 days after the urchins were released to the natural environment.

The half-life of a cohort of *Centrostephanus coronatus* tagged with stainless steel wire introduced through the test was only 15 days [22]. Duggan & Miller (2001) marked individuals of the *Strongylocentrouts droebachiensis* species with both external and internal (anchor) tags, attached to the urchin's test by means of a hole drilled in the test with the help of a needle, which caused a mortality of more than 50% of marked urchins within 1 month. Other authors who used anchor tags in the test [22, 25, 29, 46, 47] obtained a higher survival rate of urchins, although long-term survival terms in the natural environment remained low.

Passive Integrated Transponder Tags or PIT Tags are currently the most effective physical marking method in sea urchins, in relation to the uniqueness of the mark and its external readability, and do not affect the growth of urchins in the long term [30]. These labels have the advantage that they contain millions of unique codes, whereas, with chemical marking techniques such as the use of polyfluorochromes, only 4096 unique codes would be generated. Their main disadvantage is that they cannot be used to mark urchins smaller than 25 mm in diameter without causing the death of the individual.

The chemical marking method (through the use of fluorochromes) in sea urchins is also effective, also allowing the marking of a large number of urchins of any size by immersing individuals in fluorochrome baths [42]. This method is based on the incorporation of chemicals that bind to calcium, such as oxytetracycline, alizarin, calcein, etc., applied at the time of marking, which binds to the skeletal structures of various marine organisms [48]. Marking occurs through immersion, injection, or feeding.

This method has several advantages over other techniques used: 1) a large number of individuals can be marked at high speed in the natural environment; 2) minute growth increases can be detected; 3) very small individuals can be tagged by immersing them in baths containing the fluorochrome.

The main disadvantages of this technique are: 1) urchins must be slaughtered to detect the mark; 2) sample preparation is laborious and time-consuming; 3) the increase in test diameter cannot be directly measured, but is estimated from the growth increments of the individual skeletal structures; 4) in the case of skeletal resorption occurs in tagged urchins, negative growth would not be detectable.

The need to euthanize the tagged individual to detect the tag makes this tagging method unsuitable for the identification of large numbers of specimens released into the wild for repopulation purposes.

5. Minimum marking size

After verifying that the two tags that gave the best results in terms of survival and retention rate of the tag were the wires and the Trovan Minitransponders, consistent tests were carried out on marking juvenile urchins of different sizes, in a range of 10–30 mm in diameter, in order to determine the minimum size that juvenile urchins must have in order to be marked successfully without presenting mortality, obtaining a result of 13 mm in minimum diameter in the case of Trovan's Minitransponders (8 mm in length), and only 11 mm in diameter for 4–5 mm long galvanized wire sections. These minimum marking sizes are lower than those found in the literature for physical labels, which are generally not less than 20 mm in diameter without excessive mortality, and is in the range of 10 to 15 mm in length of the urchins that de la Uz et al. [49] marked with coded wire tags (CWT), allowing the monitoring of juvenile *P. lividus* individuals from 5 to 6 months of age.

6. Conclusion

The results obtained in this work allow us to conclude that Trovan's Minitransponders are an appropriate brand to monitor sea urchins of the *P. lividus* species in short and long-term studies, both in captivity and in the natural environment, presenting a high rate of retention and recapture in culture structures, and they are also suitable for marking juveniles of *P. lividus* from 13 mm of test diameter, with an approximate age of 5–6 months.

Author details

Noelia Tourón[1*], Estefanía Paredes[2,3,4] and Damián Costas[1]

1 Unit of Biological Resources and Marine Resources (UCM), Toralla Marine Sciences Station (ECIMAT), University of Vigo, Pontevedra, Spain

2 Coastal Ecology Laboratory (ECOCOST), University of Vigo, Spain

3 Department of Ecology and Animal Biology, University of Vigo, Spain

4 Marine Research Center, University of Vigo, Spain

*Address all correspondence to: noeliatbesada@yahoo.es

IntechOpen

References

[1] FAO (2016) Service de l'information et des statistiques sur les pêches et l'aquaculture. 2015. Production de l'aquaculture 1950-2013. FishStatJ-Logiciel universel pour les séries chronologiques de don- nées statistiques sur les pêches. Organisation des Nations Unies pour l'alimentation et l'agriculture. Available from: http://www.fao.org/fishery/statistics/software/fishstatj/en

[2] Espino F, Boyra A, Tuya F, Haroun RJ. Guía visual de las especies marinas de Canarias. Canary Islands (Spain): Ediciones Oceanográficas; 2006. p. 482

[3] Himmelman JH, Lavergne Y. Organization of rocky subtidal communities in the St Lawrence estuary. Naturaliste Canadien. 1985;**112**:143-154

[4] Lawrence JM. On the relationships between marine plants ans sea urchin. Oceanography and Marine Biology: An Annual Review. 1975;**13**:213-286

[5] Moro L, Martín JL, Garrido MJ, Izquierdo I. Lista de especies marinas de Canarias (algas, hongos, plantas y animales) 2003. Consejería de Política Territorial y Medio Ambiente del Gobierno de Canarias. Spain: Canary Islands Government; 2003. p. 248

[6] Vadas RL, Elner RW. Cap 2: Plant-Animal Interactions in the north-west Atlantic En Plant-Animal Interactions in the Marine Benthos. Oxford: Clarendon Press; 1992. p. 570

[7] Bachet F, Monin M, Charbonnel E, Bretton O, Cadville B. Suivi de levolution des populations d_oursins comestibles (Paracentrotus lividus) sur la Cote Bleue Resultats des comptages d_avril 2014. Rapport Parc Marin de la Cote Bleue et. France: Comite Regional des Peches Maritimes PACA; 2014. p. 17

[8] Couvray S, Miard T, Bunet R, Martin Y, Bonnefont JL, Coupe S.

Experimental release of *Paracentrotus lividus* sea urchin juveniles in exploited sites along the French Mediterranean coast. Journal of Shellfish Research. 2015;**34**(2):1-9

[9] Hereu B. Depletion of palatable algae by sea urchins and fishes in a Mediterranean subtidal community. Marine Ecology Progress Series. 2006; **313**:95-103

[10] Asnaghi V, Chiantore M, Mangialajo L, Gazeau F, Francour P, Alliouane S, et al. Cascading effects of ocean acidification in a rocky subtidal community. PLoS One. 2013;**8**:e61978

[11] FAO. FAO Food Outlook, Global Market Analysis. Rome: FAO; 2010

[12] Keesing JK, Hall KC. Review of harvests and status of world's sea urchin fisheries points to opportunities for aquaculture. Journal of Shellfish Research. 1998;**17**:1597-1604

[13] Castilla-Gavilán M, Buzin F, Cognie B, Dumay J, Turpin V, Decottignies P. Optimising microalgae diets in the sea urchin *Paracentrotus lividus* larviculture to promote aquaculture diversification. Aquaculture. 2018;**490**: 251-259

[14] Galicia, Spain. Available from: www.pescadegalicia.gal

[15] Lawrence JM, editor. Sea Urchins: Biology and Ecology. Amsterdam, The Netherlands: Elsevier B.V; 2013

[16] Lawrence JM, Agatsuma Y. Tripneustes. In: Lawrence JM, editor. Sea Urchins: Biology and Ecology. 3rd ed. Croydon, UK: Academic Press; 2013. pp. 491-507

[17] Paredes E, Bellas J, Costas D. Sea urchin (Paracentrotus lividus) larval rearing culture from cryopreserved embryos. Aquaculture. 2015;**437**:366-369

[18] Fuji AR. Studies on the biology of the sea urchin, V. Food consumption of Strongylocentrotus intermedius. Japanese Journal of Ecology. 1962;**12**:181-186

[19] Lewis GA. Geotactic movements following disturbance in the European sea urchin Echinus esculentus (Echinodermata, Echinoidea). Progress in Underwater Science. 1980;**5**:171-186

[20] McPherson BF. Contributions to the biology of Phylogeny and Selection on Echinoid Egg Size 191 the sea urchin Eucidaris tribuloides (Lamarck). Bulletin of Marine Science. 1968;**18**:400-443

[21] Neill JB. A novel technique for tagging sea urchins. Bulletin of Marine Science. 1987;**41**:92-94

[22] Nelson BV, Vance RR. Diel foraging patterns of the sea urchin Centrostephanus coronatus as a predator avoidance strategy. Marine Biology. 1979;**51**:251-258. DOI: 10.1007/BF00386805

[23] Shepherd A, Boudouresque CF. A preliminary note on the movement of the sea urchin Paracentrotus lividus. France: Scientific reports of the Port-Cros national park. 1979;**5**:155-158

[24] Cuenca C. Quelques methodes de marquages des oursins echinides (Echinodermes). Bulletin de la Societe des Sciences Naturelles de I'Ouest de la France. Nouvelle serie. 1987;**9**:26-37

[25] Dix TG. Biology of Echinus chloroticus. Echinoidea: Echinometridae fromdifferent localities. New Zealand Journal of Marine and Freshwater Research. 1970;**4**:267-277. DOI: 10.1080/00288330.1970.9515355

[26] Ebert TA. A technique for the individual marking of sea urchins. Ecology. 1965;**46**:193-194

[27] Hur SB, Yoo SK. Laboratory tagging experiment of sea urchin *Hemicentrotus pulcherrimus* (A. Agassiz). Bulletin of the Korean Fisheries Society. 1985;**18**(4): 363-368

[28] Lees DC. Tagging subtidal echinoderms. Underwater Naturalist. 1968;**5**:16-19

[29] Olson M, Newton G. A simple, rapid method for making individual sea urchins. California Fish and Game. 1979;**65**:58-62

[30] Hagen NT. Tagging sea urchins: A new technique for individual identification. Aquaculture. 1996;**139**: 271-284

[31] Kalvass PE, Hendrix JM, Law PM. Experimental analysis of 3 internal marking methods for red sea urchins. California Fish and Game. 1998;**84**:88-99

[32] Ebert TA. Relative growth of sea urchin jaws: An example of plastic resource allocation. Bulletin of Marine Science. 1980;**30**(2):467-474

[33] Ebert TA. Longevity, life history, and relative body wall size in sea urchins. Ecological Monographs. 1982;**52**(4): 353-394. DOI: 10.2307/2937351

[34] Ebert TA, Dixon JD, Schoeter SC, Kalvass PE, Richmond NT, Bradbury WA, et al. Growth and mortality of red sea urchins across a latitudinal gradient. Marine Ecology Progress Series. 1999;**190**:189-209. DOI: 10.3354/meps190189

[35] Kenner MC. Population dynamics of the sea urchin Strongylocentrotus purpuratus in a Central California kelp forest: Recruitment, mortality, growth, and diet. Marine Biology. 1992;**112**(1): 107-118. DOI: 10.1007/BF00349734

[36] Kobayashi S, Taki J. Calcification in sea urchins. Calcified Tissue Research. 1969;**4**:210-223. DOI: 10.1007/BF0 2279124

[37] Lamare MD, Mladenov PV. Modelling somatic growth in the sea urchin Evechinus chloroticus (Echinoidea: Echinometridae). Journal of Experimental Marine Biology and Ecology. 2000;**243**:17-43

[38] Pearse JS, Pearse VB. Growth zones in the echinoid skeleton. American Zoologist. 1975;**15**(3):731-751. DOI: 10.1093/icb/15.3.731

[39] Russell MP, Ebert TA, Petraitis PS. Field estimates of growth and mortality of the green sea urchin, Strongylo-centrotus droebachiensis. Ophelia. 1998;**48**:137-153

[40] Duggan RE, Miller RJ. External and internal tags for the green sea urchin. Journal of Experimental Marine Biology and Ecology. 2001;**258**:115-122. DOI: 10.1016/S0022-0981(01)00213-1

[41] Lauzon-Guay JS, Scheibling RE. Evaluation of passive integrated transponder (PIT) tags in studies of sea urchins: Caution advised. Aquatic Biology. 2008;**2**:105-112. DOI: 10.3354/ab00040

[42] Ellers O, Johnson AS. Plyfluorochrome marking slows growth only during the marking month in *Strongylocentrotus droebachiensis*. Invertebrate Biology. 2009;**128**(2): 126-144. DOI: 10.1111/j.1744-7410. 2008.00159.x

[43] Dance C. Patterns of activity of the sea urchin *Paracentrotus lividus* in the Bay of Port-Cros (Var, France, Mediterranean). Marine Ecology. 1987;**8**(2):131-142. DOI: 10.1111/j.1439-0485.1987.tb00179.x

[44] Tuya F, Martin JA, Luque A. A novel technique for tagging the long-spined sea urchin Diadema antillarum. Sarsia. 2003;**88**:365-368

[45] Lees DC. The Relationship between Movement and Available Food in the Sea Urchins Strongylocentrotus franciscanus and Strongylocentrotus purpuratus [thesis]. San Diego, California, USA: San Diego State University; 1970

[46] Hereu B. Movement patterns of the sea urchin *Paracentrotus lividus* in a marine reserve and an unprotected area in the NW Mediterranean. Marine Ecology. 2005;**26**:54-62. DOI: 10.1111/j.1439-0485.2005.00038.x

[47] James DW. Diet, movement, and covering behavior of the sea urchin Toxopneustes roseus in rhodolith beds in the Gulf of California, Mexico. Marine Biology. 2000;**137**:913-923. DOI: 10.1007/s002270000423

[48] Campana SE. Chemistry composition of fish otholits: Pathways, mechanisms and applications. Marine Ecology Progress Series. 1999;**188**:263-297

[49] de la Uz S, Carrasco JF, Rodríguez C, López J. Evaluation of tagging and substrate refuges in release of juvenile sea urchins. Regional Studies in Marine Science. 2018;**23**:8-11

Section 2

Agroecosystems

Chapter 8

Agroecological Approach to Farming for Sustainable Development: The Indian Scenario

Ishwari Singh Bisht, Jai Chand Rana, Sarah Jones, Natalia Estrada-Carmona and Rashmi Yadav

Abstract

Agroecology is the application of ecological principles to agricultural systems and practices and the application of social justice principles to whole food systems. Agroecological farming, an unfamiliar concept to those who treat agriculture and ecology as separate subjects, refers to farming for producing food, employment and economic benefits in addition to cultural, social and environmental services and benefits. Additionally, agroecology empowers farming communities, as the key agents of change, and addresses the root cause of problems of unsustainable agricultural systems in an integrated way and provides holistic and long-term solutions to transform the food and agricultural systems. As agroecology is at the forefront of transforming farming and food system sustainability, the present chapter specifically explores the state of Indian traditional farming agroecosystems, evidence collected under the ongoing Indian UNEP-GEF project "Mainstreaming agricultural biodiversity conservation and utilization in agricultural sector to ensure ecosystem services and reduce vulnerability". We discuss traditional Indian farming in view of FAO's 10 principles of Agroecology which is key to help policymakers, practitioners and stakeholders, in planning, managing and evaluating agroecological transitions.

Keywords: agroecology, agrarian reforms, traditional Indian agroecosystems, agroecological transitions, sustainable agriculture

1. Introduction

Agroecology, barely recognized a decade ago within official circles, has taken a central stage now in global discussions on food system, environment, and development [1–7]. Agroecology offers an alternative and viable strategy for transforming food systems to deliver fair outcomes for farmers, society, and the environment [8]. Its holistic view of agroecosystems facilitates ecological and social levels of coevolution, structure, and function [8]. In particular, agroecology is acquiring a new relevance on reconstructing the post-COVID-19 agriculture, one that is able to avoid widespread disruptions of food supplies in the future by territorializing food production and consumption of healthy and sustainably produced foods [9].

The Food and Agriculture Organization (FAO) of the United Nations asserts that agroecology can help alleviate hunger and poverty as well as contribute to meeting other sustainable development goals [3]. Agroecological practices such as crop diversification, intercropping, agroforestry, mixed crop-livestock systems, soil

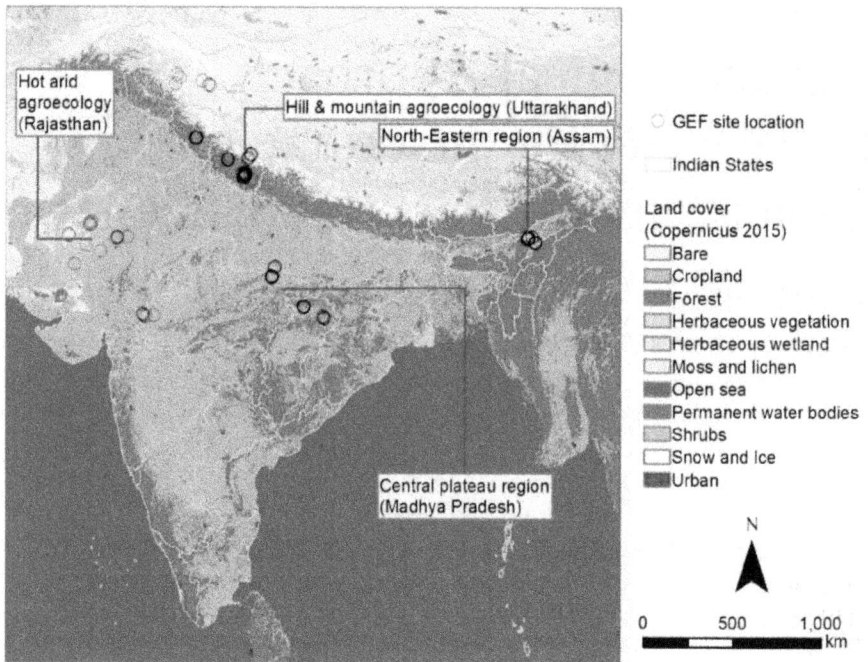

Figure 1.
The four unique agroecosystems of India being studied presently under the UNEP-GEF project and the extent of agricultural area in India.

management measures, and farmer-to-farmer networks have been reported to have positive food security and nutrition outcomes [7]. To facilitate the adoption and transition towards agroecological production systems, various efforts have helped to condense and integrate the various principles and elements put forward as key enablers of agroecology as a social movement and science [2, 6, 10, 11, 12].

Here, we use the FAO [11] approved 10 Elements of Agroecology to identify how traditional Indian farming systems contribute to each element and discuss the knowledge gaps and next steps needed to further advance adopting these principles in the Indian context. We mobilize the evidence collected from the UNEP GEF project "Mainstreaming agricultural biodiversity conservation and utilization in agricultural sector to ensure ecosystem services and reduce vulnerability". The project's evidence comes from four contrasting agroecosystems (**Figure 1**) specifically selected to cover unique crops and associated diversity adapted to diverse agricultural practices, weather pattern and socioeconomic systems. The UNEP-GEF project will directly support India's contribution to the CBD's Strategic Plan and the Aichi Targets adopted at the 10th Conference of the Parties (CoP) of the CBD, by working with farmers to document existing cultivars, and trial and scale crop diversification strategies (as one of agroecology principles) to enhance agronomic, ecological, and social outcomes [13]. Overall, we aim to offer a fundamentally different vision of the way India can produce and consume food, while contributing to the creation of equitable food systems.

2. Traditional world agriculture and the Indian agroecosystems

More than half of the world's cultivated land is still farmed by traditional and subsistence methods [11, 14]. This type of farming is usually better adapted to local conditions and has been benefitted from centuries of cultural and biological evolution. Small holder farmers have inherited or developed such complex farming systems that have helped them meet their subsistence needs for centuries, even under adverse environmental conditions with scarce and locally available resources, without depending on purchased inputs. In traditional subsistence farming, nearly all of the crops or livestock raised are used to maintain the farmer and the farmer's family, leaving little, if any, surplus for sale or trade.

The small holder farmers have designed practices that optimize productivity in the long term rather than maximize it in the short term [15]. Inputs originate locally and the farm work is performed by family labours or animals that are fuelled from local sources. Smallholder farmers, working within these energy and spatial constraints, have learned to recognize and use locally available resources for agricultural production [16]. Traditional farmers are innovative, and often manage and value a plethora of products and characteristics of the farming system (e.g. resilience, food availability), beyond the yield of only one commodity. The productivity comparisons between Green Revolution and traditional agriculture systems have, therefore, been misleading and biased. In order to remedy deficiencies in modern agriculture, many scientists in developed countries are now showing enhanced interest in traditional agriculture, especially in small-scale mixed crop systems. The wealth of traditional farmers' practical experiential knowledge needs to be transferred to farming system productions before it is lost forever.

India represents nearly 7–8% of the recorded species and is one of the 17 recognized mega-diverse countries of the world, with a large landmass and varied ecosystems [17]. It represents four of the 36 globally identified biodiversity hotspots, designated by Conservation International [18]. Besides, there are 22 recognized agrobiodiversity hotspots [19] that harbor the diversity of native and naturalized crops, their wild and weedy relatives, and crop associated biodiversity in agroecosystems.

The smallholder farming in all recognized agrobiodiversity hotspots of India is mainly subsistence and highly labour-intensive. Agriculture, however, is still the biggest land use and the biggest employer in India. Nearly 55% of the population rely on agriculture and allied activities for their livelihood [20]. Smallholder and marginal farmers account for 86.2% of all farmers in India.

3. The 10 elements of agroecology and the traditional Indian farming scenario

The 10 elements of agroecology emanated from the FAO regional seminars on agroecology and are intended to help guide countries to transform their food and agricultural system. The elements have been grouped under three categories as shown in **Figure 2** [11]. These 10 Elements of Agroecology are interlinked and interdependent, and are a guide for policymakers, practitioners and stakeholders in planning, managing and evaluating agroecological transitions, as an analytical tool.

Kumar [21] provides an insight into the agrarian history of pre-green revolution India and the reasons contributing to the adoption of "productivity-oriented" green revolution agriculture, largely unaddressed in the contemporary literature. Das [22] defines it as a tragedy that the green revolution model of agricultural development has made such headway that it is almost impossible to do away with the concept and practice of 'increasing production' of all sectors of agriculture at the cost of the environment, economy, ecology, nutrition, diversity, etc. A country with a tradition of paddy rice and millet cultivation adopted a new agricultural development strategy based mainly on wheat. The multi-crop model existing in different

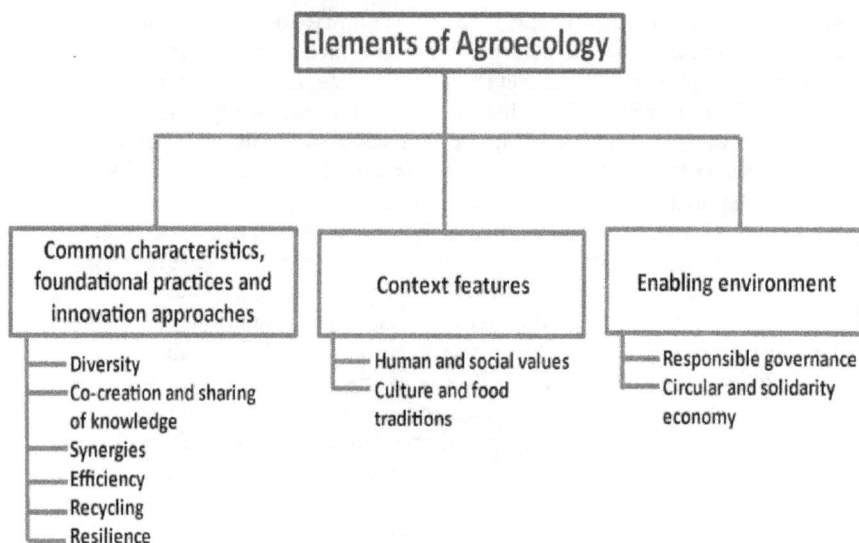

Figure 2.
The 10 elements of agroecology from FAO [11].

agroecological conditions has been neglected at the cost of environment and socio-economy, adopting a monocrop model heralded by the green revolution. The green revolution increased agricultural production around the world with the planting of mainly high yielding wheat and rice varieties that depended on applications of synthetic fertilizers, pesticides, irrigation and enhanced farm mechanization. The Green Revolution in India started in the late 1960s and with its success India attained food self-sufficiency within a decade.

The blind adherence to increasing food production without considering trade-offs or synergies with other outcomes is now being challenged [21], and enabling India to envision alternative futures that address the needs of farmers, society and nature. The self-sufficiency in two cereals, wheat and rice, in India came at the cost of another form of dependence - the import of rock phosphate for fertilizers and petroleum for irrigation pumps and tractors. Dependence on these non-renewable and fast depleting sources of energy and minerals also made agriculture a carbon-emitting sector impacting the climate. The country did not stop being vulnerable; it became vulnerable to a different set of interests. With this, deeper ecological and existential questions have emerged [22].

In the following sections, we use data collected on the UNEP-GEF project to investigate how well FAO's elements of agroecology are embedded into tra-ditional farming landscapes in the four agroecologically contrasting regions of India, described above (**Figure 1**). This includes data collected through explor-atory surveys with farmers across four representative agroecosystems based on participatory focus group discussions and observational surveys, between 2017 and 2020 [23, 24]. Farmer surveys indicate that about 80% of households have crop-livestock mixed farming across the four Indian agroecosystems, while the remaining 20% are engaged either in crop production or livestock production alone (**Table 1**). Livestock, therefore, are integral sector of all traditional farm-ing agroecosystems. Farmers indicated that the purpose of crop production is mainly for home consumption (subsistence) and only the surplus produce is for sale.

Main agricultural activity (response of mean % households)	
1. Mainly crop production	8.52
2. Mainly livestock production	10.60
3. Mixed (crop and livestock)	80.88
Purpose of crop production (response of mean % households)	
1. Producing only for sale	—
2. Producing mainly for sale with some own consumption	8.32
3. Producing mainly for own consumption with some sales	68.84
4. Producing only for own consumption	22.84
Sourced from: Bisht et al. [23].	

Table 1.
Main agricultural activity and purpose of crop production in traditional Indian farming agroecosystems.

3.1 Diversity

Biological diversity is essential to life, providing the raw material for evolution and strengthening ecological stability. This also applies to crop diversity as without it, crop improvement is impossible [25]. Traditionally, farmers worldwide, have been selecting, improving, developing, protecting and using a wide range of species adapted to the often harsh or difficult pedo-climatic conditions through ingenious practices, unfortunately, these knowledge, practices and species are disappearing fast [26]. The novel agroecosystem designs appropriate to smallholder farmers are reported to have been modeled on successful traditional farming systems [27].

Deploying and protecting currently available biodiversity in production land-scapes, contributes to a range of production, socio-economic, nutrition and environmental benefits. Diversification has been a common and key to agroecological transition ensuring food and nutrition security and sustainable management of natural resources in all the traditional Indian agroecosystems researched in the recent past [23, 24]. Traditional production systems in India are highly diverse, characterized by polyculture farming; crop-livestock small-scale mixed farming; greater farmer household production and dietary diversity; use of traditional agriculture innovation practices, etc. The benefits of diversification extend to human diets. Consuming a diverse range of food resources are important in contributing macro- and, micro-nutrients and, other bioactive compounds to human diets [28]. On-farm conservation have been reported to result in a number of interlinked elements that supports agricultural biodiversity as part of a dynamic system [29].

Multiple strategies exist to diversify production systems. For example, agro-forestry systems organize crops, shrubs, and trees of different heights and shapes at different levels or strata, increasing vertical strata with different habitat and resources [30] for biodiversity, climate mitigation and yields. Intercropping combines complementary species to increase spatial diversity. Crop rotations, often including legumes, increase temporal diversity and soil fertility by fixing nitrogen. Crop-livestock systems in all agroecosystems rely on diversity of local breeds adapted to specific environments, that also largely contribute to household cash economy and soil fertility and labour relief [31]. Similarly, traditional fish polyculture farming systems follow the same principles to maximizing diversity [32].

In the project sites, in average, traditional systems at the community level maintain three varieties per crop (**Table 2**). Surprisingly, 55% of land is occupied by rare landraces (i.e. traditional variety) despite the current lack of economic or social

Crop diversity variables	Diversity measure	Diversity estimates*
Crop species diversity	Species richness	16.0
Within- species (genetic) diversity	Cultivar richness	47.8
Area share of common landraces	Share of cropland (%)	46.0
Area share of rare landraces	Share of cropland (%)	54.0
Loss of species diversity	Species lost as a share of all known crop species** (%)	20.0
Loss of genetic (within species) diversity	Cultivars lost as a share of all known crop cultivars** (%)	7.0

*Diversity estimates were made per village, as a unit of study, based on 2–3 core villages each across four representative agroecosystems of the GEF project (Sourced from: Bisht et al. [24].
**Information on known crop species and cultivars is based on exploratory surveys.

Table 2.
Major staple food crop species and within-species (genetic) diversity in traditional farming agroecosystems.

recognition of these conservation efforts. The crop landrace species and varietal diversity remains high, although 20% of species and 7% of varieties are considered locally extinct. Crop species and varietal diversity has been maintained with the active intervention of local farmers. The traditional landraces differing in morphological characteristics, offer farmers other valued benefits including taste, texture, cooking quality, resistance to biotic/abiotic stresses and others besides yield per se. For example, in the project sites, we found each agroecosystem has a unique crop/species combination for multiple uses including food, but also medicinal, incense and perfume which have a cultural and social value (**Table 3**).

Crop diversity loss and agroecosystems homogenization have major consequences for provision of ecosystem system services as well as food system sustainability [33]. Agroecology can help reverse these trends by managing and conserving agrobiodiversity, and responding to the increasing demand for a diversity of products that are eco-friendly and nutritious. The 'fish-friendly' rice produced from rice ecosystems, particularly in tropical and subtropical Asia including India, can be cited as an example here, which values the diversity of aquatic species and their importance for rural livelihoods [34].

Conservation is especially important in the case of disappearing, specially adapted varieties, calling for renewed efforts to support farmers as custodians of biodiversity and genetic resources [35]. Hence, the importance of policies for agroecological transition that enables, recognize and strengthen the collaboration between holders of indigenous knowledge and mainstream scientific research. This close collaboration will facilitate co-producing knowledge that will guide locally relevant and adapted interventions for preserving diversity in the field, and landscapes and for food, nutrition, ecosystem services and resilience [36].

UNEP [37] outlines some of the key issues for consideration by policymakers to ensure the continued engagement of farmers in conservation and the use of agrobiodiversity. Recognizing better the role of farmers as libraries of traditional knowledge, custodians of natural resources and providers of nutritious foods and ecosystem services reflected in a better social status and quality of life could be a good starting point to encourage farmers to continue farming. This requires the support of policymakers, for developing the mix of mechanisms or incentives that will make farming an appealing, respected and well valued profession and way of

Agroecosystem	Main crops	Main tree/shrub agroforestry species
Hill & mountain	Rice (*Oryza sativa*), wheat (*Triticum aestivum*), minor millets (ragi, *Eleusine coracana*; barnyard millet, *Echinochloa frumentacea*); foxtail millet, *Setaria italica*), black-seeded soybean (*Glycine max*), urd bean (*Vigna mungo*), horsegram (*Macrotyloma uniflorum*), mustard (*Brassica* spp.), sesame (*Sesamum indicum*), pseudocereals (amaranths, *Amaranthus* spp.; buckwheat, *Fagopyrum* spp.), miscellaneous vegetables, temperate fruits, etc.	Main agroforestry species for high quality fiber are drooping Fig (*Ficus semicordata*), *Grewia oppositifolia*, *G. asiatica* etc., and for edible fruits are European nettle tree (*Celtis australis*,) *Grewia oppositifolia, G. asiatica*, Elephant ear Fig (*Ficus auriculata*), wild Fig (*F. palmata*), drooping fig (*F. semicordata*), willow-leaf fig (*F. nemoralis*), wild Himalayan pear (*Pyrus pashia*), etc., beside several others.
Hot arid	Pearl millet (*Pennisetum glaucum*), mung bean (*Vigna radiata*), sesame (*Sesamum indicum*) and cluster bean (*Cyamopsis tetragonaloba*)	Screw bean (*Prosopis cineraria*), *Ziziphus nummularia,* wild Caper bush (*Capparis decidua*), gum arabic tree (*Acacia senegal*).
Central tribal plateau	Rice (*Oryza sativa*), wheat (*Triticum aestivum*), pigeon pea (*Cajanus cajan*), mung bean (*Vigna radiata*), urd bean, soybean (*V. mungo*)	Forestry species: gum arabic tree (*Acacia nilotica*), river tamarind (*Leucaena leucocephala*), English beechwood (*Gmelina arboria*), North Indian rosewood (*Dalbergia sissoo, Pongame oiltree (Millettia pinnata*), and as fruit trees: Malabar plum (*Syzygium cumini*), common guava (*Psidium guajava*), drumstick tree (*Moringa oleifera*), Indian gooseberry (*Phyllanthus emblica*), custard apple (*Annona reticulata*), jackfruit (*Artocarpus heterophyllus*).

Agroecosystem	Main crops	Main tree/shrub agroforestry species
North-eastern region	Rice (*Oryza sativa*), tea (*Camellia sinensis*), vegetables, sugarcane (*Saccharum officinarum*), jute (*Corchorus olitorius*), cotton (*Gossypium* spp.), black gram (*V. mungo*), lentil (*Lens culinaris*) green gram (*V. radiata*), gram (*Cicer arietinum*), pigeon pea (*Cajanus cajan*), linseed (*Linum usitatissimum*), castor (*Ricinus communis*), sesame (*Sesamum indicum*), rapeseed & mustard (*Brassica* spp.), banana (*Musa* spp.), papaya (*Carica papaya*), orange (*Citrus* spp.), pineapple (*Ananas comosus*), areca nut (*Areca catechu*), coconut (*Cocos nucifera*), chili (*Capsicum* spp.), turmeric (*Curcuma longa*), ginger (*Zingiber officinale*), potato (*Solanum tuberosum*), sweet potato (*Ipomoea batatas*), etc.	Agar (*Aquilaria agallocha*), areca nut (*Areca catechu*), needlewood tree (*Schima wallichii*), java cassia (*Cassia nodosa*), kassod tree (*Cassia seamea*), white siris (*Albizzia procera*), betel (*Piper betel*), long pepper (*P. longum*), bamboos (*Bambusa* spp.), canes, timbers and other shade trees.

Sourced from: Bisht et al. [24].

Table 3.
Main agroforestry species of the different Indian farming agroecosystems and multiple uses (fibers, food, fodder, medicinal, wood, incense/perfume).

living. For example, payment for ecosystem services (PES) can compensate farmers for the services and conservation efforts they provide, beyond food. Similarly, market signals that makes cheaper traditional and nutritious food, against highly subsidized and less nutritious food can also incentivize the production, consumption and profitability of traditional/indigenous crops, this off course should be in tandem with nutritional programs that highlight the nutritional, ecological, agricultural and cultural value of traditional foods. Investing in conservation, protection and use of agrobiodiversity in field and plates is an urgent need across countries for enabling and facilitating agroecological transitions and production systems that provide nutritious food and ecosystem services. Investing, therefore, where the most agricultural biodiversity occurs, subsistence farming, is an important low-risk option.

3.2 Co-creation and sharing of knowledge

In traditional Indian farming contexts, we find limited responsiveness of modern science to societal needs [24]. The gap between experts' knowledge and traditional innovations in actual farming situations were more pronounced when sustainability issues are being considered. Sustainability of traditional smallholder farming, therefore, requires a holistic approach and an interdisciplinary research style. The need of a new knowledge base has been strongly felt for transition towards more sustainable agriculture [38]. Farmers greatly value local experiential knowledge which is not being optimally used and a better strategy to integrate various forms of knowledge is needed [39].

Incorporating farmers' experiential knowledge with formal agricultural knowledge is still being debated [40], as the agricultural knowledge system has always been very closely connected to the modernisation process in agriculture,

Management areas and various management actions based on farmers' indigenous knowledge (IK)
• **Biodiversity conservation**
○ Management of domesticated and wild farm biodiversity
○ Local community-level on-farm and off-farm vegetation management including forestry resources
○ Managing biodiversity in sacred groves/sacred landscapes
○ Cultivation of medicinal plants.
• **Adaptation to climate change**
○ The multiple and diversified livelihood skills of farmers is a source of resilience in times of uncertain weather and climate change.
○ Maintaining species and genetic diversity in fields provide a low-risk buffer in uncertain weather and the diversity in production landscapes is considered a necessity rather than a choice.
• **Agroforestry**
○ Indigenous knowledge on traditional agroforestry offers opportunities to farmers for sustainable management of resources and support socio-ecological and socio-economic benefits.
○ The traditional/cultural knowledge embedded within the rural communities in different agroecosystems is the inherent identity that is unique and diverse in all respects to traditional agroforestry management and conservation. It is reflected in their cultivation system, ethnobiology and health and nutrition management.
• **Traditional medicine**
○ Use of herbal medicines was reported by native farming communities of all Indian agroecosystems. Traditional medicines are used to cure different ailments. Herbal formulations were administered either internally or applied externally depending on the type of ailment.
• **Customary resource management**
○ Traditional knowledge, innovations and practices duly supported by spiritual beliefs and customary laws are developed and nurtured over many generations. The natural resource-based livelihood of native communities enable them to live within the natural limits of specific territories, areas or resources upon which they depend for livelihoods and wellbeing.
• **Applied anthropology**
○ Indigenous knowledge and institutions are contributing to more culturally appropriate and sustainable development. It is also based on the realization that native communities are not only more keenly aware of their needs than are outside development agencies but that those needs are culturally defined, demanding a substantive rather than a formal appreciation.
• **Impact assessment**
○ Indigenous knowledge can assist bring awareness about potential impact of a project and steps taken to prevent adverse effects to the existing environment but there are currently no guidelines on how indigenous knowledge should be integrated into impact assessments.
• **Natural disaster preparedness and response**
○ Indigenous knowledge can be transferred and adapted to other communities in disaster management, it encourages community participation and empowers communities in reducing disaster risk.
Sourced from: Bisht et al. [24].

Table 4.
Farmers' experiential knowledge and various management actions related to mainstreaming biodiversity in production landscapes.

the 'scientification' of agriculture [41]. The science-based model advocating yield maximization, for example, often fail in actual farming situations and farmers normally find that experts' knowledge is of limited practical value [42–44]. This gap between theory and practice becomes even more pronounced when sustainability issues need to be considered and calls for a new mode of working that enables scientists to optimize knowledge within and for different local conditions. In order for agriculture to become sustainable and resilient, there is need of knowledge networking that facilitates knowledge exchanges, joint learning which facilitates the generation and innovation of new and more integrated solutions [39].

Agricultural innovations respond better to local challenges when they are co-created through participatory processes. Data on farmers' experiential knowledge and various management actions related to mainstreaming biodiversity across the four UNEP-GEF project production landscapes are presented in **Table 4**, collected through participatory focus group discussion meetings with farmers from 2 to 3 core villages in each of the four representative agroecosystems. This shows that in traditional systems farmers' experiential knowledge to agriculture cannot be seen in isolation, rather a whole range of interlinked management areas are as important.

Native farming communities in all Indian agroecosystems are especially vulnerable to weather uncertainties and climate change [24]. The community level climate change adaptation plans are often rooted in Western scientific knowledge, largely ignoring traditional farmer innovations. Incorporating indigenous knowledge into Western science-based climate change adaptation plans is, therefore, an untapped opportunity for the policymakers to integrate into climate change adaptation plans and legislate accordingly.

Farmers' knowledge is considered a better resource for managing ecosystems [45] that gives an insight on designing social systems that mesh better with ecosystems. The differential farming styles are forms of adapting to diversity within local ecosystems. Farming styles are an outcome of 'co-production', that is the ongoing interplay and mutual transformation of the social and the technical [46], including evidently local ecosystems.

As agroecology depends on context-specific knowledge, hence agroecological practices should be tailored to fit the environmental, social, economic, cultural and political context [11]. The co-creation and sharing of knowledge at multiple levels (i.e. farmers, states, ecoregions, countries) plays a central role in the process of developing and implementing agroecological innovations to address challenges across food systems including adaptation to climate change. Currently, media can facilitate fast and massive knowledge interchange with a larger reach than traditional extension officers. Hence, co-evaluating agroecological practices (i.e. extension officers, universities, research institutions and farmers) across agroecological zones and social-economic context through simple online videos and tutorials verified and curated is a new strategy at its infancy for facilitating knowledge integration and farmer-to-farmer learning at the new pace and scale that is needed. Through the co-creation process, agroecology cross-pollinate traditional knowledge and global scientific knowledge.

3.3 Synergies

The IAASTD [47] concluded that the future of agriculture lies in biodiverse, agroecological based farming systems that can meet social, economic and environmental goals while maintaining and increasing productivity. Agroecology is therefore increasingly recognized as the way forward for agriculture, capable of delivering productivity goals without depleting the environment and disempowering communities. The value of various ecosystem services to agriculture is enormous and often underappreciated [48, 49]. Agroecosystems also produce a variety of ecosystem services, such as conservation of biodiversity, regulation of soil and water quality, carbon sequestration, and cultural services [50]. On the other hand, depending on management practices, agriculture can also be the source of numerous disservices, including loss of wildlife habitat, nutrient runoff, sedimentation of waterways, greenhouse gas emissions, and pesticide poisoning of humans and non-target species [51].

The trade-offs that may occur between provisioning services and other ecosystem services and disservices should be evaluated in terms of spatial scale, temporal scale and reversibility. As more effective methods for valuing ecosystem services become available, the potential for 'win-win' scenarios increases. Under all scenarios, appropriate agricultural management practices are critical to realizing the benefits of ecosystem services and reducing disservices from agricultural activities [47]. Building synergies enhances key functions across food systems, supporting production and multiple ecosystem services.

Agroecology pays careful attention to the design of diversified systems that selectively combine annual and perennial crops, livestock, trees, soils, water and other components on farms and agricultural landscapes to enhance synergies in the context of an increasingly changing climate [11].

Building synergies in food systems delivers multiple benefits. By optimizing biological synergies, agroecological practices enhance ecological functions, leading to greater resource-use efficiency and resilience. Intercropping with pulses in traditional farming landscapes saves about USD 10 million in nitrogen fertilizers globally every year through biological nitrogen fixation [52] and substantially contributes to soil health, climate change mitigation and adaptation. Crop-livestock integration in traditional farming systems also highlights synergies as about 15 percent of the nitrogen applied to crops comes from livestock manure [53]. Integrated rice systems, in Asia, combine rice cultivation with the generation of other products such as fish, ducks and trees. The total area of land available for rice cultivation in India is about 43 million hectares (ha), of which an estimated 20 million ha is suitable for adoption of the rice-fish integration system, mainly in rainfed medium lands, waterlogged lands etc. By maximizing synergies, integrated rice systems significantly improve yields, dietary diversity, weed control, soil structure and fertility, as well as providing biodiversity habitat and pest control [54].

At the landscape level, synchronization of productive activities in time and space is necessary to enhance synergies between social and nature rhythms. Pastoralism and extensive livestock grazing systems manage complex interactions between people, multi-species herds and variable environmental conditions, building resilience and contributing to ecosystem services such as seed dispersal, habitat preservation and soil fertility [55, 56]. In India, the beauty of pastoralist ways of life lies in their ability to convert the marginal resources in dry and arid regions; cold mountain meadows, and other regions to productive resources such as milk, meat, wool, and manure with marginal inputs.

While agroecological approaches strive to maximize synergies, trade-offs also occur in natural and human systems. Managing trade-offs is an endless process and innate characteristic of sustainable production systems, hence the urgent need of getting good at anticipating, managing and meditating trade-offs. Agroecology emphasizes the importance of partnerships, cooperation and responsible governance, involving different actors at multiple scales to promote synergies within the wider food system, and best manage trade-offs.

In India, a good example of actively seeking for synergies is the agroecological system of farming millets, in water deficient Tamil Nadu [57]. They ought to take into account not only water use, but also the whole gamut of political and ecological

Agricultural inputs used in traditional farming agroecosystems*	
Use of farmer varieties or traditional landraces (%)	80.5
Use of purchased inputs (%)	
• Seeds	11.3
• Inorganic Fertilizer	6.3
• Pesticides	—
Use of improved mechanized modern farming practices (%)	10.0
Area share of crops that have non-food uses (%)	6.3

Percent of households in a village, as a unit of study. In total 2–3 villages each in four representative agroecosystems of the GEF project sites were surveyed.
Sourced from: Bisht et al. [24].

Table 5.
Characteristics of inputs used in traditional Indian farming agro-ecosystems.

issues that are connected to farming such as public procurement, land reform, minimum support price, subsidized credit, agricultural extension services, and so on. The publicly procured millet output is distributed through the public distribution system, government schools, and through the network of Amma canteens in the state. Amma canteens is a food subsidization programme, a first of its kind scheme, run originally by the Government of Tamil Nadu state in India. Its success has been an inspiration for many other states of India including Odisha, Karnataka and Andhra Pradesh, who subsequently proposed similar schemes.

3.4 Efficiency

Agroecological systems improve the use of natural resources, especially those that are abundant and free, such as solar radiation, atmospheric carbon and nitrogen. Innovative agroecological practices produce more using less external resources. **Table 5** lists the inputs used in traditional Indian farming landscapes from across the UNEP-GEF study sites. Use of external resources or purchased inputs is minimal in traditional agroecosystems.

Increased resource-use efficiency is an emergent property of agroecological systems that carefully plan and manage diversity to create synergies between different system components. A key efficiency challenge, for example, is that less than 50 percent of nitrogen fertilizer added globally to cropland is converted into harvested products and the rest is lost to the environment causing major environmental problems [11, 58].

By enhancing biological processes and recycling biomass, nutrients and water, traditional farming communities are able to use fewer external resources, reducing costs and the negative environmental impacts of their use. Reducing dependency on external resources will ultimately empowers farmers by increasing their autonomy and resilience to natural or economic shocks. Agroecology thus promotes agricultural systems with the necessary biological, socio-economic and institutional diversity and alignment in time and space to support greater efficiency. Nonetheless, technological, social and digital innovations remain a critical need for offering farmers timely information and reducing labour constraints in already heavily overworked farmers.

3.5 Recycling

More recycling means agricultural production with lower economic, social and environmental costs. In all traditional subsistence Indian agroecosystems, farming is organic by default with negligible use of inorganic fertilizers and pesticides. Organic farming is primarily aimed at cultivating the land and raising crops in such a way, as to keep the soil alive and in good health by use of organic wastes and avoiding use of synthetic inputs (such as inorganic fertilizers, pesticides, hormones, feed additives etc.). The organic cultivation rely to the maximum extent feasible upon crop rotations, crop residues, animal manures, forest litter and other off-farm organic waste, and biological system of nutrient mobilization and plant protection. The default organic agriculture in all Indian agroecosystems is a unique production management system which promotes and enhances agroecosystem health, including biodiversity, biological cycles and soil biological activity, and this is accomplished by using on-farm agronomic, biological and mechanical methods in exclusion of all synthetic off-farm inputs.

Recycling can take place at both farm-scale and within landscapes, through diversification and building of synergies between different components and activities. Agroforestry systems, for example, that include deep rooting trees can capture nutrients lost beyond the roots of annual crops [59].

Crop–livestock systems of all traditional Indian agroecosystems, promote recycling of organic materials by using manure for composting or directly as fertilizer, and crop residues and by-products as livestock feed. Integrating livestock plays a key role in nutrient cycling, accounting substantially of the economic value of all non-provisioning ecosystem services. Recycling organic materials and by-products offers great potential for agroecological innovations.

3.6 Resilience

Enhanced resilience of farmers, communities and ecosystems is key to sustainable food and agricultural systems. It is a well-accepted fact now that the high external input agriculture of India, with the spread of green revolution paradigm and technology, is unsustainable and has placed enormous strain on natural resource base of the economy. India's agrarian issues are being discussed widely in policy circles and media but the solutions proposed by policy makers hardly seem to be addressing the deep structural malaise that has set in at the core of India's agrarian economy [60]. In the north-western India, the core green revolution area is experiencing, massive groundwater depletion, high land degradation, decline in the levels of soil organic matter, soil erosion, loss of soil fertility without mentioning the countless impacts on human health and wellbeing. By revisiting India's agrarian history and outlining the circumstances under which the green revolution model was adopted, Kumar [21] has sought to challenge the blind adherence of high productivity agriculture that are likely to open up ways to address the future needs of the country.

The search for solutions to problems that plague Indian agriculture must begin with fundamentally questioning the green revolution paradigm. Taking agroecology as its core, the alternative path calls for making a decisive shift from a production/economic-centric approach of the green revolution paradigm towards an ecosystem/social-centric approach for resilient farming system.

High crop species and varietal (within-species or genetic) diversity provides resilience by contributing to production stability and by enabling long-term adaptation. Intercropping of varieties with varying water-use efficiencies stabilized yield in a drought-prone environment. Adaptation to long-term environmental changes may reflect both phenotypic plasticity and continued evolution [61]. Maintaining evolutionary processes ensures the generation of new combinations of genes in response to stresses and climatic variability [62]. Cultivation of varietal mixtures may confer enhanced resilience to biotic and abiotic stresses [63, 64]. Crop yield increased by 2.2 percent overall were reported in cultivar mixtures with more functional-trait diversity in comparison to monoculture in a study examining the relationship between intraspecific diversity and yield in cultivar mixtures [65]. Cultivar mixtures also showed higher yield stability than monocultures, especially in response to annual weather variability over time.

Enabling agroecological adoption in India requires, therefore, promoting and safeguarding the free access of farmers to a range of diverse varieties will improve the resilience of production systems [66] by maintaining and supporting seed-exchange systems, and associated traditional knowledge. Farmers' traditional knowledge, preferences and practices, and social networks strongly influence the stress-prone and marginal production systems by enhanced use of genetic resources [67–69]. Informal seed-exchange systems are especially effective at maintaining high diversity, and participation in social networks has been demonstrated to facilitate access to genetic resources that can aid farmers in coping with crop failures, drought and environmental uncertainties [70]. On a landscape scale, diversified agricultural landscapes have a greater potential to contribute to pest and disease control functions [71].

3.7 Human and social values

Protecting and improving rural livelihoods, equity and social well-being is essential for sustainable food and agricultural systems. Across diverse settings, the traditional agricultural "living landscapes", created by native peoples and local communities are the results of the dynamic interaction of people and nature over time. These landscapes, rich in agrobiodiversity as well as inherent wild biodiversity and cultural and spiritual values, embody human ingenuity and are continually evolving [72]. These landscapes and their associated management systems have much to teach us about sustainability and resilience in the face of global change.

Agroecology places a strong emphasis on human and social values. Agroecological approaches empower communities to overcome poverty, hunger and malnutrition by building autonomy and adaptive capacities to manage their agroecosystems. Agroecological approaches also promote human rights, such as the right to good and healthy food, and stewardship of the environment so that future generations can also live in prosperity.

Agroecology also seeks to address gender inequalities by creating opportunities for women or other minorities often left behind or ignored. Women make up almost half of the agricultural workforce in India. Besides household food security, nutrition and health, women also play a vital role in conservation and sustainable use of biological diversity. Their contribution, however, remain unrecognized making them economically marginalized and vulnerable to violations of their rights [73].

Table 6 highlights the role of women as agricultural workforce and contribution of women in household cash income in different farming agroecosystems. Self-help groups (SHG) is bringing women together under a common platform. In addition to their farming skills, women are learning stitching, embroidery, patchwork, weaving, food-processing (making use of their locally available resources), handicrafts, etc. Women enjoy the learning opportunity as well as the quality time spend together. Working for few hours a day for certain days (7–10) in a month, the women associated with the group are earning a decent amount, which represents in average about 10% of the household cash income. Participating in the SHG has helped them to gain self-respect and increase their say in decision making of family matters. Agroecology can, therefore, help open spaces for women to become more autonomous and empower them at household, community levels and beyond – for instance, through participation in producer groups. Women's participation is essential for agroecology and women are frequently the leaders of agroecology projects.

Agroecosystems	Agriculture workforce		Contribution of women SHG to HH cash income (%)
	Men (%)	Women (%)	
Hill & mountain agroecology (Uttarakhand)	36	62	15
Arid desert (Rajasthan)	56	78	7
Central plateau region (Madhya Pradesh)	54	80	8
North-eastern region (Assam)	51	72	12
Mean	49.3 (±9.0)	73.0 (±8.1)	10.5 (±3.7)

Sourced from: Bisht et al. [24].

Table 6.
Agriculture workforce in different agroecosystems and contribution of women self-help groups (SHGs) to household (HH) cash income.

In many places around the world, rural youth face a crisis of employment. Agroecology provides a promising solution as a source of decent jobs. Agroecology is based on a different way of agricultural production that is knowledge intensive, environmentally friendly, socially responsible, innovative, and which depends on skilled labour. Meanwhile, rural youth around the world possess energy, creativity and a desire to positively change their world. What they need is stable, good and long-term support and opportunities.

Table 7 indicates probable areas where enhanced job opportunities at community level exist in small-holder Indian farming. As a bottom-up, grassroots paradigm for sustainable rural development, agroecology empowers people to become their own agents of change.

3.8 Culture and food traditions

By supporting healthy, diversified and culturally appropriate diets, agroecology contributes to food security and nutrition while maintaining the health of ecosystems.

The food we eat plays a huge role in our ability to keep our physical, mental, emotional and psychological health in balance. Further, it is being greatly recognized now that without mental health there can be no true physical health. In spite of nutrition transition trends, it is widely acknowledged that the traditional farming agroecosystems rich in crop and livestock diversity and use of wild harvested foods, the food traditions are still prevailing in the life of rural households to a greater extent. This is indeed heartening that the traditional food habits are still playing a great role in contemporary food habits of the traditional Indian farming communities; therefore, the possibility of reversing the trends in favor of dietary diversification from dietary simplification looks promising with enabling policies [74].

Culture and food traditions play a central role in society and in shaping human behavior as agriculture and food are core components of the human heritage. However, in many instances, our current food systems have created a disconnection between food habits and culture. This disconnection has contributed to a situation where hunger and obesity exist side by side, in a world that produces enough food to feed its entire population. Almost 800 million people worldwide are chronically hungry and 2 billion suffer micronutrient deficiencies [75]. Meanwhile, there has been a rampant rise in obesity and diet-related diseases; 1.9 billion people are overweight or obese and non-communicable diseases (cancer, cardiovascular disease, diabetes) are the number one cause of global mortality, a pattern also followed in India [76, 77].

Probable areas	Job opportunities and policy support required
Organic farming	• Production of organic agricultural inputs. • Post-harvest farm - to - market supply chains. • Linking organic farming to marketing interventions. • Infrastructure development like cold stores to avoid post-harvest losses.
Agro-ecotourism	• Linking ecotourism to traditional farming landscapes. • Developing herbal farms, food parks, biodiversity parks, sacred grooves, fish farms, wild life parks, rural game parks in agricultural landscapes near to ecotourism sites. • Training the local youths in hospitality management and environmental education.
Women-centric jobs, viz. embroidery, tailoring, weaving, patchwork, applique, handicraft, etc.	• Creating women-centric jobs by forming Self Help Groups (SHGs). • Requisite skill development and making available all need-based equipment/resources at subsidized rates.
Management of Common Property Resources (CPRs)/ agroforestry species/ community forests	• Nursery raising and planting of agroforestry species. • Planting diverse tree species and maintaining diverse economically important species at CPRs .

Sourced from: Bisht et al. [23].

Table 7.
Probable areas/sectors where job opportunities at community level exist in small-holder Indian farming.

Agroecology seeks to cultivate a healthy relationship between people, culture and food by rebalancing traditions and modern food habits. Cultural identity and sense of place are often closely tied to landscapes and food systems. As people and ecosystems have evolved together, cultural practices and indigenous and traditional knowledge offer a wealth of experience that can inspire agroecological solutions. For example, India is home to an estimated 50 000 indigenous varieties of rice [78]- bred over centuries for their specific taste, nutrition and pest-resistance properties, and their adaptability to a range of conditions. Culinary traditions are built around these different varieties, making use of their different properties. Taking this accumulated and tasty body of traditional knowledge as a guide, agroecology can help realize the potential of territories to sustain their peoples.

3.9 Responsible governance

Agroecology calls for responsible and effective governance to support the transition to sustainable food and agricultural systems. **Table 8** indicates areas where largely default organic production of traditional Indian farming agroecosystems can be liked to "localized" marketing interventions through enabling marketing support. Transparent, accountable and inclusive governance mechanisms are necessary to create an enabling environment that supports producers to transform their systems following agroecological concepts and practices. Promoting community supported agriculture (CSA); linking traditional farming with school feeding (MDM) and public procurement programmes; market regulations allowing for branding of differentiated agroecological produce, and subsidies and incentives for ecosystem services, etc. are some areas where a strong political will and policy support is required for sustainable farming and food systems.

Linking organic farming to market-oriented initiatives	Actions at community level and policy support
Community Supported Agriculture (CSA) initiatives	• Facilitate forming village-level farmer cooperatives.
	• Consolidation and pooling of farm land for collective farming.
	• Awareness campaigns for popularizing the nutritional superiority of organically grown native crops among urban consumers.
	• Mobilizing urban consumers become CSA members.
Linking organic food to school meal (MDM) programmes	• Empowering the local district administrations to make changes to menu of MDM served in government schools to suit the local tastes.
	• Divert a substantial Ministry of Human Resource Development (MHRD) budget for local purchasing of native healthy food for MDM directly from small-holder native farmers.
Enhanced market access and value chain development for local plant food resources	• Make food-based approach as major initiative for household nutrition and health.
	• Where there is no secure market for raw produce, build capacity of farmers for processing/packaging to enhance benefit from localized sale.

Sourced from: Bisht et al. [23].

Table 8.
Linking organic farming to marketing interventions, community level actions and policy support.

Promoting Community-Supported Agriculture (CSA) Initiatives: In small holder Indian farming, CSA initiatives are considered a better approach for sustainable agricultural development [23]. Hence, CSA can gain and play a larger role on satisfying the local/regional nutritional and food needs, where consumers are becoming more concerned about the environment, health, and animal welfare, due to food scandals related to the industrialized and globalized food systems [79–81]. Also, alternative production systems such as organic agriculture, are distancing from its original ideology due to the high industrialization and long supply chains [82, 83]. Therefore, new and alternative arrangements bringing farmers and consumers closer together and shortening supply chains, like farmers' markets, farm shops, subscription box schemes, and CSA are supporting sustainable farming and consumption [84, 85].

Reinforcing solidarity, direct human relationships, mutual trust, small scale, and respect for the environment and life overall is key in the new economic and sustainable models [86]. Most CSAs were initially based on vegetable production but a wide range of other agricultural produce is increasingly being covered now [87]. Therefore, the agroecological transition should strengthen and bring closer farmers and consumers to return to both production and food sovereignty, where farmers and consumers are empowered, autonomous, happy and where farmer's key role in culture and society valued and recognized [88].

Linking traditional agriculture with school meal (MDM) programme: A crucial step in enhancing the nutritional standards of MDM in traditional farming agroecosystems of India, would be through the introduction of nutritionally rich local crops in the MDM menu [23].

Structured demand guarantees large yet predictable sources of demand to smallholder or marginal farmers, thereby giving them income security. A cooperative model would be better as farmers would retain bargaining power in the supply chain. As for local schools in villages, the cooperatives themselves could supply the commodities based on the requirement, which is how it works in Brazil [89]. However, there is need of ensuring strict monitoring at every stage of procurement and payment. For a school meal scheme to be a success for both children and farmers, like Brazil's PNAE, existing structural loopholes in both the education and agricultural sectors have to be plugged [90]. Eventually, such contracts could also be extended from school meals to include public colleges, offices and hospitals.

Enhanced market access and value chain development for local plant food resources: Enhanced "localized" market access and value chain development for local plant food resources can be an important initiative, making traditional agriculture in Indian agroecosystems sustainable [23]. The native crops from traditional farming areas have a greater potential for value chain development and other marketing interventions. There is enough scope for development of local and distant markets in which traditional varieties command a price premium. With enhanced awareness about the nutritional importance of local crops in the community, in well-functioning markets, the native crop landraces can be competitive and have enough potential to provide commercial opportunities fetching a premium price (**Table 9**).

Off-farm employment for rural youth at community level: Farm and non-farm employment opportunities at community level for rural youths is considered very vital to bring sustainability in agricultural production [23]. Policies that help to generate part-time, farm and non-farm employment at community level in rural areas can, therefore, help sustain small farms. Organic farming; agro-ecotourism; women-centric self-help groups (SHGs) for several non-farm jobs viz. embroidery, tailoring, weaving, patchwork, applique, handicraft, etc., and community managed agroforestry/forestry interventions can generate enough jobs for rural youths for year-round employment. Non-farm income already accounts for a significant proportion of household income in rural India [91]. Hence, the Mahatma Gandhi National Rural Employment Guarantee Act (MNREGA) of India, aimed at enhancing livelihood security in rural areas at the community level for reducing out-migration of rural youth in search of off-farm employment elsewhere, but the scheme often failed due to misappropriation and subversion of funds in many states [24].

Promoting food-based approach towards community nutrition and health: Food-based approach towards community nutrition and health under overall eco-nutrition framework, needs to be promoted [92–94]. An econutrition model has been suggested for a healthy human nutrition that can be best achieved by an

Agroecology	Organic crops with high marketing potential
Hill & mountain	Common bean (*Phaseolus vulgaris*), soybean (local black-seeded, *Glycine max*), black gram (*Vigna mungo*), horse gram (*Macrotyloma uniflorum*), finger millet (*Eleusine coracana*), barnyard millet (*Echinochloa frumentacea*), buckwheat (*Fagopyrum* spp.), amaranths (*Amaranthus* spp.), aromatic (including *basmati*) and red rice (*Oryza sativa*).
Hot arid	Coriander (*Coriandrum sativum*), fennel (*Foeniculum vulgare*), fenugreek (*Trigonella foenum-graceum*), mung bean (*Vigna radiata*), pearl millet (*Pennisetum glaucum*), sesame (*Sesamum indicum*).
Central tribal plateau	Sharbati wheat (*Triticum aetivum*); durum wheat (*Triticum durum*), pigeonpea (*Cajanus cajan*); Kodo-Kutki (*Paspalum scrobiculatum* and *Panicum sumatrense*); Basmati/aromatic rice (*Oryza sativa*), and organic cotton grown in different parts of central plateau region are in great demand nationally/ internationally.
North-eastern region	Joha (aromatic) rice (*Oryza sativa*), ginger (*Zingiber officinale*), turmeric (*Curcuma longa*), chili (*Capsicum* spp.), oranges (*Citrus* spp.), black pepper (*Piper nigrum*) and pineapples (*Ananas comosus*).

Sourced from: Bisht et al. [24].

Table 9.
Organically grown crops with high marketing potential grown in some Indian agroecosystems.

approach to agriculture that is biodiverse. Integrating environmental and human health, focusing especially on the many interactions between agriculture, ecology, and human nutrition are being explored [95, 96]. A more radical transformation of agriculture will be required for development of sustainable agriculture by ensuring that ecological change in agriculture is only possible with comparable changes in the social, political, cultural, and economic arenas that help determine agriculture.

An inter-disciplinary collaboration is required to define priority research questions to co-deliver economic, environmental and health goals [97]. Food-based solutions to hunger, malnutrition and poverty are of global concern and must be addressed if food and nutrition security is to be achieved in a sustainable manner [98]. According to the HLPE [99], "A sustainable food system (SFS) is a food system that delivers food security and nutrition for all in such a way that the economic, social and environmental bases to generate food security and nutrition for future generations are not compromised".

Promoting indigenous food sovereignty: Food sovereignty prioritizes local and national economies and markets and empowers peasant and family farmer-driven agriculture, artisanal fishing, pastoralist-led grazing, and food production, distribution and consumption based on environmental, social and economic sustainability [100].

In traditional Indian agroecosystems, without any formal interventions, food sovereignty exists de facto. Re-introduction of indigenous food production practices will help restore food sovereignty to native communities [101]. The food sovereignty initiatives, world over, are community-led. There are reports that many tribal communities in USA, for example, are regaining control of their food supply, they are growing traditional foods and collaborating with the federal government to retain rights for hunting and gathering [102].

The subsistence farming agroecosystems of India are expected to set the stage for future research that demonstrates how the local foods contribute to a sustainable agriculture–food–nutrition strategy. In India, there is no formal awareness about indigenous food sovereignty movements and no formal partnerships with native farming communities doing their part to address the challenges linked to ensuring indigenous food sovereignty. Formal Food Sovereignty Alliances need to put the traditional farming communities at the centre of decision-making on policies, strategies and natural resource management [103, 104].

3.10 Circular and solidarity economy

Circular economy is based on the principles of eliminating waste, continued use of resources and regenerating natural systems [105]. The solidarity economy refers to a wide range of economic activities that prioritizes social profitability over purely financial profits.

As per a recent tentative estimate by an Environmental Research and Action Group "CHINTAN" (*www.chintan-india.org/sites/default/files/2019–09/ Food%20 waste%20in% 20 India.pdf*), about 40% food produced in India is wasted. Despite adequate food production, it has been reported by the UN that about 190 million Indians remain undernourished. It is further estimated that the value of food wastage in India is around ₹92,000 crores (13,000 million USD) per annum. These are some bleak statistics, but they help us realize the magnitude of the problem of food waste, as much as inequity, in India. A substantial food waste along the food chain, accounting for more than 30% of the agricultural production, is also a big concern at the global level [106].

Prioritizing local markets and supporting local economic developments by creating virtuous cycles, agroecology seeks to reconnect producers and consumers through a circular and solidarity economy [107]. Agroecological approaches help create more equitable and sustainable markets by promoting fair solutions based on local needs, resources and capacities. Shorter food circuits strengthen better producer-consumer linkages. It can increase the incomes of food producers while maintaining a fair price for consumers [107]. These include new innovative markets [108, 109] alongside more traditional territorial markets, where most smallholders market their products.

Regenerative agriculture is already gaining momentum in India. Application of circular economy principles is likely to make agricultural production more regenerative, creating a more diverse and resilient food system; preserving the integrity of the natural systems, and supporting rural livelihoods and incomes [110].

A broad range of circular economy opportunities exists for India to consider when shaping the future of its food system and agricultural activities until 2050. By capturing these opportunities, India could build a food and agricultural system that leverages the current small-farm structure to create a network of farmers, symbiotic in their practices and committed to regenerative approaches.

With on-going COVID-19 pandemic, it has forced us to revisit the way we tend to modify our agricultural practices. We need to sustain out food and nutritional security, farming with new technological developments vis-à-vis traditional agroecological methods need to be merged. In short, learnings from the past need to be married with the present practices while eyeing the future.

Kumbamu [111] critically examines and analyses place-based as well as network-based strategies of alternative development organizations that claim to be building sustainable social and solidarity economies (SSE) in the political context of neoliberal globalization. While the Indian state and market forces are actively promoting the neoliberal agri-food system, alternative development organizations are working with farmers to build the SSE based on the principles of democracy, inclusiveness, reciprocity, cooperativism, and socioecological sustainability. Using a case study approach, the article analyses how SSE initiatives are aiming to reclaim control over the local agri-food sector. Specifically, the article examines how community development organizations mobilize farmers based on the principles of agroecology and the politics of seed and food sovereignties.

4. Conclusion

Agroecosystem management is at a crossroads today. The challenge that modern agriculture is presently facing is not to increase productivity but to strengthen the resilience of our food production in the face of ever increasing stress on the system. The major long-term trends and challenges faced, as highlighted in the FAO [75] report, will determine the future of food security and nutrition, rural poverty, the efficiency of food systems, and the sustainability and resilience of rural livelihoods, agricultural systems and their natural resource base.

With the present global climate change and the dwindling natural resource base, it will be difficult to continue growing food in a way that will support the future human generations. This is where agroecology comes in, which is the foundation of sustainable agriculture and the best path forward for feeding the world. Agroecology provides robust set of solutions to the environmental and economic problems for design and management of sustainable farms.

The agroecological approach specifically aims to transform agriculture to build locally relevant food systems that strengthen the economic viability of rural areas based on short marketing chains, and both fair and safe food production.

Increasing dependence on hazardous pesticides and other purchased chemical inputs will degrade soil, pollute water and threaten the essential ecological services to agriculture. By shifting farming policies and practice to embrace agroecology, we can create a food system - one rooted in productivity, resilience, equity and sustainability to sustain this and future human generations.

Traditional farming practices as adopted by small holder Indian farming communities in different agroecosystems showcase a better way forward that recognizes multifunctional dimensions of agroecological approaches to agriculture including use of locally available resources and indigenous knowledge and practices. The de facto organic biodiverse agriculture of different traditional Indian production landscapes demonstrates a bottom-up, grassroots paradigm for sustainable rural development empowering native farmers to become their own agents of change. This has power to protect and improve rural livelihoods, equity and social well-being, considered essential for sustainable food and agricultural systems.

As agroecology requires a whole-systems approach based on traditional knowledge, alternative agriculture and local food system experiences, blending modern science with farmers' experiential knowledge is considered important for enabling agricultural innovations. Agricultural innovations respond better to local challenges when they are co-created through participatory processes. The local food movement and year-round employment opportunities for rural youth are key components to help bring about the much needed transformation to agroecological farming and food systems. Making agriculture more environmentally, socially and economically sustainable will, in turn, lead to overall rural development, critical to shifting India's rural farmers out of poverty.

Acknowledgements

The authors thank the project partners and the farmer households of different farming agroecosystems for effectively interacting and sharing valuable information on traditional farming and native food systems under the UNEP-GEF project "Mainstreaming agricultural biodiversity conservation and utilization in agricultural sector to ensure ecosystem services and reduce vulnerability in India (Project Code: A 1265)' being executed jointly by Alliance of Bioversity International and CIAT, and the Indian Council of Agricultural Research (ICAR).

Conflicts of interest

The authors declare no conflict of interest.

Author details

Ishwari Singh Bisht[1*], Jai Chand Rana[1*], Sarah Jones[2], Natalia Estrada-Carmona[2] and Rashmi Yadav[3]

1 The Alliance of Bioversity International and CIAT, NASC Complex, Pusa Campus, New Delhi, India

2 The Alliance of Bioversity International and CIAT, Montpellier, France

3 ICAR-National Bureau of Plant Genetic Resources, Pusa Campus, New Delhi, India

*Address all correspondence to: bisht.ishwari@gmail.com and j.rana@cgiar.org

IntechOpen

References

[1] Caron P, de Loma-Osorio GFy, Nabarro D, Hainzelin E, Guillou M, Andersen I, Arnold T, Astralaga M, Beukeboom M, Bickersteth S, Bwalya M, Caballero P, Campbell BM, Divine N, Fan S, Frick M, Friis A, Gallagher M, Halkin JP, Hanson C, Lasbennes F, Ribera T, Rockstrom J, Schuepbach M, Steer A, Tutwiler N, Verburg G. Food systems for sustainable development: Proposals for a profound four-part transformation. Agronomy for Sustainable Development. 2018; 38:41. https://doi.org/10.1007/s13593-018-0519-1

[2] Gliessman S. Defining Agroecology. Agroecology and Sustainable Food Systems. 2018; 42(6): 599-600, DOI: 10.1080/21683565.2018.1432329

[3] Pimbert MP. Global Status of Agroecology: A Perspective on Current Practices, Potential and Challenges. Economic and Political Weekly. 2018; 53(41): 52-57.

[4] Anderson CR, Pimbert MP, Chappell MJ, Brem-Wilson J, Claeys P, Kiss C, Maughan C, Milgroom J, McAllister G, Moeller N, Singh J. Agroecology now - connecting the dots to enable agroecology transformations, Agroecology and Sustainable Food Systems, 2020, 44 (5): 561-565, DOI: 10.1080/21683565.2019.1709320

[5] Barrios E, Gemmill-Herren B, Bicksler A, Siliprandi E, Brathwaite R, Moller S, Batello C, Tittonell P. The 10 Elements of Agroecology: enabling transitions towards sustainable agriculture and food systems through visual narratives, Ecosystems and People. 2020;16 (1): 230-247, DOI: 10.1080/26395916.2020.1808705

[6] Wezel A, Herren BG, Kerr RB, Barrios E, Gonçalves ALR, Sinclair F. Agroecological principles and elements and their implications for transitioning to sustainable food systems. A review. Agronomy for Sustainable Development. 2020; 40:40. https://doi.org/10.1007/s13593- 20-00646-z.

[7] Kerr RB, Madsen S, Stüber M, Liebert J, Enloe S, Borghino N, Parros P, Mutyambai MD, Prudhon M, Wezel A. Can agroecology improve food security and nutrition? A review. Global Food Security. 2021; 29:100540. https://doi.org/10.1016/j.gfs.2021.100540

[8] Altieri MA. Agroecology: The Science of Sustainable Agriculture. CRC Press. Taylor & Francis Group, Boca Raton, FL), 1995.

[9] Altieri MA, Nicholls CI. Agroecology and the reconstruction of a post-COVID-19 agriculture, The Journal of Peasant Studies. 2020; 47: 881-898. DOI: 10.1080/03066150.2020.1782891Alti eri MA. Agroecology: The Science of Sustainable Agriculture. CRC Press. Taylor & Francis Group, Boca Raton, FL), 1995.

[10] Gliessman SR. Agroecology: The Ecology of Sustainable Food Systems, Third Edition. Boca Raton, FL, USA, CRC Press, Taylor & Francis Group), 2015.

[11] FAO. The 10 elements of agroecology: Guiding the transition to sustainable food and agricultural systems. The Food and Agriculture Organization, Rome Italy. ISBN: I9037EN/1/04.18, 2018. (http://www.fao.org/documents/card/en/c/I9037EN

[12] HLPE. Agroecological and other innovative approaches for sustainable agriculture and food systems that enhance food security and nutrition. A report by the High Level Panel of Experts on Food Security and Nutrition of the Committee on World Food Security, Rome. 2019. http://www.fao.org/cfs/cfs-hlpe/en

[13] CBD (Convention on Biological Diversity). Strategic plan for biodiversity 2011-2020. Aichi Biodiversity Targets, 2018.

[14] Ruthenberg H. Farming Systems of the Tropics. London: Oxford Univ. Press, 1971.

[15] Gliessman SRE, Garda R, Amador AM. The ecological basis for the application of traditional agricultural technology in the management of tropical agro-ecosystems. Agro—ecosystems. 1981; 7: 173-185.

[16] Wilken GC. Integrating forest and small-scale farm systems in middle America. Agro-ecosystems. 1977; 3: 291-302.

[17] Subramanian A. Biodiversity profile of India. Technical Report. 2017. DOI: 10.13140/RG.2.2.10664.57601.

[18] Biodiversity Hotspots; https//www.conservation.org.

[19] Nayar, M.P.; Singh, A.K.; Nayar, K.N. Agrobiodiversity Hotspots in India: Conservation and Benefit Sharing. Protection of Plant Varieties and Farmers' Rights Authority Government of India, New Delhi, 2009.

[20] Agricultural Census 2015-2016. All India Report on Number and Area of Operational Holdings; DAC&FW, Ministry of Agriculture & Family Welfare, GoI: New Delhi, India, 2018.

[21] Kumar R. India's green revolution and beyond: Visioning agrarian futures on selective readings of agrarian pasts. Economic and Political Weekly. 2019; 54(34): 41-48.

[22] Das SK. Pitfalls of Green Revolution. Economic and Political Weekly. 2019; 54(37).

[23] Bisht IS, Rana JC, Ahlawat SP. The future of smallholder farming in India:

Some sustainability considerations. Sustainability, 2020; 12: 3751. DOI:10.3390/su12093751

[24] Bisht IS, Rana JC, Yadav R, Ahlawat SP. Mainstreaming agricultural biodiversity in traditional production landscapes for sustainable development: The Indian scenario. Sustainability. 2020; 12: 10690; DOI:10.3390/su122410690

[25] Long J, Cromwell E, Gold K. On-farm management of crop diversity: an introductory bibliography. London: Overseas Development Institute for ITDG. 2000. 42 p.

[26] FAO. The State of the World's Biodiversity for Food and Agriculture, J. Bélanger & D. Pilling (eds.). FAO Commission on Genetic Resources for Food and Agriculture Assessments. Rome. 2019; 572 pp. (http://www.fao.org/3/CA3129EN/CA3129EN.pdf)

[27] Altieri MA. Linking Ecologists and Traditional Farmers in the Search for Sustainable Agriculture. Frontiers in Ecology and the Environment. 2004; 2: 35-42.

[28] FAO. Sustainable Diets and Biodiversity: Directions and Solutions for Policy, Research and Action. 2010. Rome.

[29] ECPGR. ECPGR Concept for on-farm conservation and management of plant genetic resources for food and agriculture. European Cooperative Programme for Plant Genetic Resources, Rome, Italy. 2017.

[30] Prabhu R, Barrios, E, Bayala J, Diby L, Donovan J, Gyau A, Graudal L, Jamnadass R, Kahia J, Kehlenbeck K, Kindt R, Kouame C, McMullin S, van Noordwijk M, Shepherd K, Sinclair F, Vaas, P, Vågen TG, Xu J. Agroforestry: realizing the promise of an agroecological approach. In: FAO. Agroecology for Food Security and

Nutrition: Proceedings of the FAO International Symposium, 2015; pp. 201-224. Rome.

[31] FAO. Ecosystem Services Provided by Livestock Species and Breeds, with Special Consideration to the Contributions of Small-Scale Livestock Keepers and Pastoralists. Commission on Genetic Resources for Food and Agriculture Background Study. 2014; Paper No. 66, Rev. 1 (available at: www.fao.org/3/aat598e.pdf).

[32] Ridler N, Wowchuk M, Robinson B, Barrington K, Chopin T, Robinson S, Page F, Reid G, Szemerda M, Sewuster J, BoyneTravis S. Integrated Multi – Trophic Aquaculture (IMTA): A potential strategic choice for farmers. Aquaculture Economics & Management, 2007;11: 99-110.

[33] Jackson LE, Pascual U, Hodgkin T. Utilizing and Conserving Agrobiodiversity in Agricultural Landscapes. Agriculture, Ecosystems & Environment. 2007;121: 196-210.

[34] Halwart M, Bartley DM. Aquatic biodiversity in rice-based ecosystems. In: Jarvis D, Padoch C, Cooper D, editors. Managing biodiversity in agricultural ecosystems. British Columbia Press .2007. p. 181-199.

[35] Gemmill B. Managing agricultural resources for biodiversity Conservation. A guide to best practices. UNEP/UNDP Biodiversity Planning Support Programme. Environment Liaison Centre International Nairobi, Kenya, 2001.

[36] Nakashima DJ; Galloway MK, Thulstrup HD, Ramos CA, Rubis JT. Weathering Uncertainty: Traditional Knowledge for Climate Change Assessment and Adaptation. Paris, UNESCO, and Darwin, UNU, 2012, 100 p.

[37] UNEP. Farmers and the future of agrobiodiversity. COP 9 MOP 4, Bonn, Germany. 2008.

[38] Stuiver M, Leeuwis C; Ploeg JD van der. The power of experience: Farmer's knowledge and sustainable innovations in agriculture. In: Wiskerske JSC, Ploeg JD van der, editors. Seeds of transition: Essays on novelty production, niches, and regimes in agriculture, Assen: Royal van Gorcum. 2004; 93-117.

[39] Šumane S, Kunda I, Knickel K, Strauss A, Tisenkopfs T, Rios I des los, Ashkenazy A. Local and farmers' knowledge matters! How integrating informal and formal knowledge enhances sustainable and resilient agriculture. J. Rural Stud., 2018; 59: 232-241. DOI: 10.1016/j.jrurstud. 2017.01.020

[40] Leeuwis, C. Learning to be sustainable. Does the Dutch agrarian knowledge market fail? European Journal of Agricultural Education and Extension. 2000; 7: 79-92

[41] Ploeg JD van der. The scientification of agricultural activities Wageningen LU. 1987. 336p (in Dutch)

[42] Scoones I, Thompson J. Beyond Farmer First: rural people's knowledge, agricultural research and extension practice. London, Intermediate Technology Publications. 1994.

[43] Eshuis J, Stuiver M, Verhoeven F, Ploeg JD van der. Good manure does not stink: a study on slurry manure, experiential knowledge and reducing nutrient losses in dairy farming. Studies van Landbouw en Platteland No 31, Circle for Rural European Studies, Wageningen University, Wageningen. 2001. 138 p. (In Dutch).

[44] Timmer WJ. Agricultural science, a philosophical essay about agriculture and agricultural science as a basis of renewal for agricultural higher education, Buitenzorg. 1949. 306p (in Dutch.

[45] Toledo VM. The ecological rationality of peasant production.

In: Altieri MA, Hecht SB, editors. Agroecology and Small Farm Development. Boston, CRC Press. 1990. p. 53-60.

[46] Law J. Power, action and belief, a new sociology of knowledge? London, Routledge. 1986. 280 p.

[47] IAASTD. Agriculture at a crossroads. International assessment of agricultural knowledge, Science and technology for development, Sub-Saharan Africa (SSA) Report. ISBN 978-1-59726-538-6. 2009.

[48] Brander L, Brouwer R, Wagtendonk, A. Economic valuation of regulating services provided by wetlands in agricultural landscapes: A meta-analysis. *Ecol. Eng.* 2013; **56**:89-96.

[49] Teixeira-Duarte G, Santos PM, Cornelissen TG, Ribeiro MC, Paglia AP. The effects of landscape patterns on ecosystem services: meta-analyses of landscape services. *Landsc. Ecol.* 2018; **33**:1247-1257.

[50] Teixeira-Duarte G, Santo, PM, Cornelissen TG, Ribeiro MC, Paglia A P. The effects of landscape patterns on ecosystem services: meta-analyses of landscape services. *Landsc. Ecol.* 2018; **33**: 1247-1257.

[51] Lade SJ, Steffen W, de Vries W, Carpenter SR, Donges JF, Hoff H, Newbold T, Richardson K, Rockström J. Human impacts on planetary boundaries amplified by Earth system interactions. *Nat. Sustain.* 2020; **3**:119-128 (2020)].

[52] FAO. Soils and Pulses: Symbiosis for life. Rome. 2016.

[53] FAO. 2017. Sustainable Agriculture for Biodiversity – Biodiversity for Sustainable Agriculture. Rome.

[54] FAO. Scaling-up integrated rice-fish systems – Tapping ancient Chinese know-how. South–South Cooperation. 2016. (available at: www.fao.org/3/a-i4289e. pdf.

[55] FAO. Ecosystem Services Provided by Livestock Species and Breeds, with Special Consideration to the Contributions of Small-Scale Livestock Keepers and Pastoralists. Commission on Genetic Resources for Food and Agriculture Background Study Paper No. 66, Rev. 1 (available at: www.fao. org/3/a@598e.pdf). 2014.

[56] Krätli S, Shareika N. Living off uncertainty: the intelligent animal production of dryland pastoralists. Eur. J. Dev. Res. 2010; 22: 605-622.

[57] Saratchand C. Agroecological Farming in Water-deficient Tamil Nadu. Economic and Political Weekly. 2018; 53 (41).

[58] Ladha JK, Pathak H, Krupnik TJ, Six J, van Kessel C. Efficiency of fertilizer nitrogen in cereal production: retrospects and prospects. Advances in Agronomy, 2005; 87: 85-156.

[59] Buresh RJ, Rowe EC, Livesley SJ, Cadisch G, Mafongoya P. Opportunities for capture of deep soil nutrients, In van Noordwijk, M., Cadisch, G., Ong, C.K. (eds.), Belowground Interactions in Tropical Agroecosystems, CAB International, Wallingford (UK). 2004, p. 109-125.

[60] Vijayshankar PS. Towards a Resilient Farming System, Economic and Political Weekly. 2019; 54 (26-27).

[61] Buresh R, Rowe EC, Livesley SJ, Cadisch G, Mafongoya P. Opportunities for capture of deep soil nutrients. In: van Noordwijk M, Cadisch G, Ong CK, editors. Below ground Interactions in Tropical Agroecosystems, CAB International, Wallingford (UK). 2004. p. 109-125.

[62] Bellon MR. Do we need crop landraces for the future? Realizing the

global option value of in situ conservation. In: Kontoleon A, Pascual U, Smale M, editors. Agrobiodiversity and Economic Development. Routledge, London and New York. 2009. p. 51-61.

[63] Bellon MR, Dulloo E, Sardos J, Thormann I, Burdon JJ. In situ conservation— harnessing natural and human-derived evolutionary forces to ensure future crop adaptation. Evolutionary Applications, 2017; 10: 965-977. https://doi.org/10.1111/eva.12521

[64] Jarvis D, Hodgkin T, Bhuwon S, Fadda C, Lopez-Noriega I. An heuristic framework for identifying multiple ways of supporting the conservation and use of traditional crop varieties within the agricultural production systems. Critical Reviews in Plant Sciences, 2011; 30: 125-176.

[65] Reiss ER, Drinkwater LE. Cultivar mixtures: a meta-analysis of the effect of intraspecific diversity on crop yield. Ecological Applications, 2017; 28: 62-77. https://doi.org/10.1002/eap.1629.

[66] Creissen HE, Jorgensen TH, Brown JK. Increased yield stability of field-grown winter barley (*Hordeum vulgare* L.) varietal mixtures through ecological processes. Crop Protection, 2016; 85: 1-8. https://doi.org/10.1016/j.cropro.2016.03.001.

[67] Brush SB. Farmers' bounty: locating crop diversity in the contemporary world. New Haven, USA, Yale University Press. 2004.

[68] Brush SB. Genes in the field: on-farm conservation of crop diversity. Rome, International Plant Genetic Resources Institute; Ottawa, International Development Research Centre; and Boca Raton, USA, Lewis Publishers. 2000.

[69] Labeyrie V, Bernard R, Leclerc, C. How social organization shapes crop diversity: an ecological anthropology approach among Tharaka farmers of Mount Kenya. Agriculture and Human Values, 2014: 31: 97-107. DOI:10.1007/s10460-013-9451-9

[70] Pautasso M, Aistara G, Barnaud A, Caillon S, Clouvel P, Coomes O, Deletre M. et al. Seed exchange networks for agrobiodiversity conservation. A review. Agronomy for Sustainable Development, 2013; 33: 151-175.

[71] Perfecto I, Vandermeer J. The agroecological matrix as alternative to the land-sparing/agriculture intensification model. Proceedings of the Natural Academy of Sciences, 2010; 107: 5786-5791

[72] Brown J, Kothari A. Traditional agricultural landscapes and community conserved areas: an overview. Management of Environmental Quality: An International Journal, 2011; 22: 139-153. DOI 10.1108/14777831 111113347.

[73] FAO & Asian Development Bank. Gender equality and food security— women's empowerment as a tool against hunger. ADB: Mandaluyong City, Philippines. 2013.

[74] Bisht IS, Mehta PS, Negi KS, Verma SK, Tyagi RK, Garkoti SC. Farmers' rights, local food systems and sustainable household dietary diversification: A case of Uttarakhand Himalaya in north-western India. Agroecology and Sustainable Food Systems, 2018; 42:73-113. DOI:10.1080/2 1683565.2017.1363118

[75] FAO. The future of food and agriculture – Trends and challenges. Rome. 2017.

[76] WHO. Obesity and overweight. 2015. (available at: www.who.int/mediacentre/ factsheets/fs311/en/).

[77] Whitmee S, Haines A, Beyrer C, Frederick B, Capon AG, de Souza Dias BF, Ezeh A, Frumkin H, Gong P, Head P, Horton R, Mace GM, Marten R, Myers SS, Nishtar S, Osofsky SA, Pattanayak SK, Pongsiri MJ, Romanelli C, Soucat A, Vega J, Yach D. Safeguarding human health in the Anthropocene epoch: report of The Rockefeller Foundation – Lancet Commission on planetary health. *Lancet. 2015;* **386**:1973-2028. https://www.thelancet.com/journals/langlo/article/PIIS2214-109X%2818%2930448-0/fulltext]

[78] NBPGR [National Bureau of Plant Genetic Resources (ICAR)]. Why do we conserve plant genetic resources? 2013. (available at: www.nbpgr.ernet.in).

[79] Renting H, Marsden TK, Banks J. Understanding alternative food networks: Exploring the role of short food supply chains in rural development. Environ. Plan. 2003; 35:393-411.

[80] Terragni L, Torjusen H, Vittersø G. The dynamics of alternative food consumption: Contexts, opportunities and transformations. In: Terragni L, Boström M, Halkier B, Mäkelä J, editors. Anthropology of food, Can consumers save the world? 2009. Retrieved from http://aof.revues.org/index6400.htm.

[81] Lamine C, Bellon S. Conversion to organic farming: A multidimensional research object at the crossroads of agricultural and social sciences. A review. Agron. Sustain. Dev. 2009; 29:97-112.

[82] Feagan R. The place of food: Mapping out the 'local' in local food systems. Prog. Hum. Geogr. 2007; 31: 23-42.

[83] Torjusen H, Lieblein G, Vittersø G. Learning, communicating and eating in local food-systems: The case of organic box schemes in Denmark and Norway. Local Environ. 2008; 13:219-234

[84] Hvitsand C. Community supported agriculture (CSA) as a transformational act—Distinct values and multiple motivations among farmers and consumers. Agroecol. Sustain. Food Syst. 2016; 40:333-351.

[85] Gliessman S. Agroecology and social transformation. Agroecol. Sustain. Food Syst. 2014; 38: 1125-1126

[86] Weckenbrock P, Volz P, Parot J, Cressot N. Introduction to Community Supported Agriculture in Europe. In: European CSA Research Group: Overview of Community Supported Agriculture in Europe; FAO: Rome, Italy. 2016. p. 8-11. Available online: http://urgenci.net/the-csa-research-group

[87] European CSA Research Group. Overview of Community Supported Agriculture in Europe; FAO: Rome, Italy. 2016. Available online: http://urgenci.net/the-csa-research-group/).

[88] Patel R. Grassroots voices: What does food sovereignty look like? J. Peasant Stud. 2009; 36:663-706.

[89] Kitaoka K. The National School Meal Program in Brazil: A Literature Review. The Japanese Journal of Nutrition and Dietetics. 2018; 76(Supplement):S115-S125. DOI:10.5264/eiyogakuzashi.76.S115.

[90] What India can learn from Brazil to ramp up its fight against hunger and poverty (https://yourstory.com 2016/10/brazil-hunger-poverty-learnings/amp)

[91] Lanjouw P, Shariff A. Rural non-farm employment in India: Access, incomes and poverty impact. Economic and Political Weekly, 2004; 39: 4429-4446.

[92] Bisht IS. Food-based approaches towards community nutrition and health: A case of Uttarakhand hills in North-Western India. Journal of Food Science and Toxicology, 2018; 2:5.

[93] Bisht IS. Globalization of food choices negatively impacting sustainability of traditional food systems: A case of Uttarakhand hills in north-western India. American Journal of Food and Nutrition, 2019; 7: 94-106. DOI:10.12691/ajfn-7-3-4

[94] Bisht IS, Pandravada SR, Rana JC, Malik SK, Singh A, Singh PB, Ahmed F, Bansal KC. Subsistence farming, agrobiodiversity and sustainable agriculture: A case study. Agroecology and Sustainable Food Systems, 2014; 38:890-912. DOI:10.1080/21683565.20 14.901273.

[95] Blaslbalg TL, Wispelwey B, Deckelbaum RJ. Econutrition and utilization of food-based approaches for nutritional health. Food Nutr. Bull. **2011**; 32 (Suppl. 1): S4–S13. DOI:10.1177 /15648265110321S102

[96] Heywood VH. Overview of agricultural biodiversity and its contribution to nutrition and health. In: Jessica F, Hunter D, Borelli T, Mattei F, editors. Routledge, Taylor & Francis Group: London, UK; New York, NY, USA. Diversifying Food and Diets: Using Agricultural Biodiversity to Improve Nutrition and Health. 2013. p 35-67

[97] Gill M, Feliciano D, Macdiarmid J, Smith P. The environmental impact of nutrition transition in three case study countries. Food Secur. 2015; 7: 493-504.

[98] Thompson B, Amoroso L (eds.). Improving Diets and Nutrition-Food-Based Approaches; The Food and Agriculture Organization of the United Nations and CABI: Rome, Italy, 2014

[99] High Level Panel of Experts on Food Security and Nutrition (HLPE). Food Losses and Waste in the Context of Sustainable Food Systems; A report by the High-Level Panel of Experts on Food Security and Nutrition of the

Committee on World Food Security: Rome, Italy, 2014

[100] Nyeleni. Declaration of Nyeleni. 2007. Available online: https://nyeleni. org/IMG/pdf/DeclNyeleni-en.pdf

[101] NICOA (National Indian Council on Aging). The Importance of Food Sovereignty; NICOA, Albuquerque, USA, 2019. Available online: https:// www.nicoa.org/the-importance-of-food-sovereignty.

[102] Native Diabetes Wellness Program. Traditional Foods in Native America: A Compendium of Stories from the Indigenous Food Sovereignty Movement in American Indian and Alaska Native Communities; Native Diabetes Wellness Program, Centres for Disease Control & Prevention: Atlanta, GA, USA, 2013

[103] Bye BAL. Native Food Systems Organizations: Strengthening Sovereignty and (Re)Building Community. Master's Thesis, Iowa State University, Ames, IA, USA, Graduate Theses and Dissertations 11121, 2009. Available online: https://lib.dr.iastate. edu/etd/11121

[104] Gliessman S. Confronting Covid-19 with agroecology. Agroecol. Sustain. Food Syst. 2020; 44:1115-1117. DOI:10.1 080/21683565.2020.1791489.

[105] Ellen MacArthur Foundation: The circular economy in detail. https://www. ellenmacarthurfoundation.org/explore/ the-circular-economy-in-detail

[106] FAO. Food Loss and Food Waste. Food and Agriculture Organization of the United Nations. 2011. (http://www. fao.org/food-loss-food-waste/flw-data)

[107] FAO. Agroecology knowledge hub. www.fao.org/agroecology/knowledge/ 10-elements/circular-economy/en/

[108] FAO/INRA. Innovative markets for sustainable agriculture – How innovations in market institutions

encourage sustainable agriculture in developing countries. Rome. 2016.

[109] FAO/INRA. Constructing markets for agroecology – An analysis of diverse options for marketing products from agroecology. Rome. 2018.

[110] Ellen MacArthur Foundation. Circular Economy in India: Rethinking growth for long-term prosperity. 2016, http://www.ellenmacarthurfoundation. org/ publications/).

[111] Kumbamu A. Building sustainable social and solidarity economies: Place-based and network-based strategies of alternative development organizations in India. Community Development. 2017; 49(1):1-16. DOI:10.1080/15575330. 2017.13844.

Chapter 9

The Role of Woody Plant Functional Traits for Sustainable Soil Management in the Agroforestry System of Ethiopia

Hana Tamrat Gebirehiwot, Alemayehu Abera Kedanu and Megersa Tafesse Adugna

Abstract

A woody plant functional trait that directly affects its fitness and environment is decisive to ensure the success of an Agroforestry practice. Hence, recognizing the woody plant functional traits is very important to boost and sustain the productivity of the system when different plants are sharing common resources, like in Agroforestry system. Therefore, the objective of this paper was to understand how woody plant functional traits contribute to sustainable soil management in Agroforestry system and to give the way forward in the case of Ethiopia. The contribution of woody plant species in improving soil fertility and controlling soil erosion is attributed by litter accumulation rate and the season, decomposability and nutrient content of the litter, root physical and chemical trait, and spread canopy structure functional trait. However, spread canopy structure functional trait is used in coffee based Agroforestry system, while with management in Parkland Agro forestry System. Woody species of Agroforestry system added a significant amount of soil TN, OC, Av.P, K, Na, Ca, and Mg nutrients to the soil. Woody plant species of Agroforestry system and their functional traits are very important to ensure sustainable soil management. Thus, further investigation of the woody plant functional traits especially the compatibility of trees with cops is needed to fully utilize the potential of woody species for sustainable soil management practice.

Keywords: woody plant, functional trait, sustainability, soil fertility, soil erosion, agroforestry system

1. Introduction

Woody plant functional trait, morphological-physiological-phenological characters that measured at an individual level and directly affects its fitness [1] and environment [2] is decisive to ensure the success of the Agroforestry practice. Agroforestry is indicated to be a prominent strategy to address land degradation, food security, and climate change challenges in Africa in general and in Ethiopia in particular too [3]. This is due to Agroforestry is a dynamic, ecologically based, natural resource management system that, through the integration of woody plants in farm- and rangeland, diversifies and sustains smallholder production for increased environmental, economic and social benefits [4].

Land degradation is a common environmental problem in Ethiopia for many years back to date due to the natural capital of the land resource is declining from time to time [5–7]. The primary reason of land degradation is land exploration for agricultural purpose to feed the ever increasing population [5, 6, 8] and predicted to be continued with the current trend [9]. As a result, soil erosion, droughts, loss of biodiversity and food insecurity are challenging the daily life of the rural population in Ethiopia [10, 11].

Therefore, integrated land use system such as Agroforestry system is very essential to combat land degradation and its consequences like soil degradation to ensure sustainable use of resources [12]. Hence, recognizing the woody plant functional traits are very important to boost and sustain the productivity of the system when different plants are sharing common resources like in Agroforestry system [13]. The canopy feature of woody plants, the height, diameter, specific root length and leaf area are among others refer to the morphological traits while the internal process and chemical composition of the woody plant denotes to physiological traits of the plant [14]. Phenology of woody plant species defines the timing of different phases of life cycle such as leaf shading and re-growing, flowering, fruiting and seed dispersal [15, 16]. Thus, these functional traits of woody plants in Agroforestry system are the core feature in supporting sustainable soil management. Wherein sustainable soil management refers to an optimum level of field soil health and productive capacity to provide ecosystem services such as provision of clean water, hydrologic and nutrient cycling, habitats for microorganisms and mesofauna, carbon sequestration, and climate regulation [17].

Hence, Agroforestry systems provide different ecosystem services. Different researchers confirmed that Agroforestry systems in Ethiopia endowed with highly diversified woody species [18–20]. The woody species diversity in Agroforestry system have indispensable role of natural forest conservation [21] as the farmers use woods from the Agroforestry system than natural forest. Moreover, the Agroforestry practices are central for keeping biodiversity and soil fertility at levels which are similar to the natural forest [22]. Research from Southern Tigray in Ethiopia indicates that Agroforestry practice has decreased soil erosion of the area [23]. However, how woody plants' functional traits support sustainable soil management has not been explored and reported in detail. Thus, in this review, how woody plants' functional traits support sustainable soil management in the Agroforestry system are discussed from soil fertility improvement and soil erosion control perspectives.

2. Woody plant functional traits and sustainable soil management

2.1 Above ground woody plant trait

Woody plants improve soil fertility and control soil erosion through their litter, canopy and root systems [24]. Ref. [25] holds a similar opinion when he states that from litter perspectives, 100% of the respondent from Jabithenan District, North-Western Ethiopia confirmed that Home garden Agroforestry system produce higher litter stock from weeds, grasses, and tree leaves than non-tree system. A similar finding by [26] reveals that in West Guji Zone, South Ethiopia, farmers noted that tree species that sheds its leave before the onset of rain and can easily decomposed are integrated in to farm land to increase the soil fertility. For example, *Faidherbia albida* and *Ziziphus mucronata* tree species are special fertilizer trees used to integrate on cropland in the parkland Agroforestry system [27]. These trees shed their leaves before the onset of rain and regrow their leaves and flower during dry season [28]. Ref. [29] state that majority of trees contribute leaf litter fall to the system in the dry season. Ref. [30] holds a similar opinion when he states that litter accumulation rate and season of the leaves fall, the decomposability of woody plant leaf are very important aspects of functional trait for improving soil fertility of the system.

Regarding to canopy, [31] state the shape of the canopy of the woody plants and the size of the leaves are very crucial in minimizing soil erosion rate. Their major findings reveal that in the case of Bonga and Yayu-Hurumu districts, Southwestern Ethiopia, 98.2% of respondents preferred woody plants with thin and small leaves in decreasing the intensity of soil erosion than broader and larger leaves as coffee shade. Additionally, spreading canopy nature of woody plants can reduce the energy of raindrops by intercepting rainfall than narrow one. Hence, protecting the soil surface against the impact of rainfall drops by intercepting runoff [32].

However, [28] states that trees such as *Cordia africana*, *Croton macrostachyus*, *Acacia etbaica*, *Ficus thoninngii*, *Sesbania sesban* and *Leucena leucocephala* has an adverse shade effect on crop production and managed through pruning to be used on agricultural land. In the parkland Agroforestry system, trees like *Cordia africana*, *Croton macrostachyus*, and *Acacia tortilis* have generally a negative effect on total aboveground biomass and grain yields of Maize in the case of Meki and Bako, Ethiopia [33]. Ref. [34] show that wheat is the most compatible crop when integrated with *Acacia albida* under shade conditions followed by maize, while teff is highly susceptible to shading effect pointing the importance of lopping to minimize shade effect. Similarly, conventional agriculture with trees highly reduces crop yield in the equatorial savannah of East Africa [35].

Concerning tree phenology of leaf fall and flowering period of woody plants, [36] indicated that farmers especially women, have limited knowledge. Their finding also reveal that farmers have better knowledge on fruiting time of edible fruit tree species because it is related to their income generation for Lemo District in SNNPR Region.

2.2 Below ground woody plant trait (root trait)

From root morphological trait perspective, uses of mixture of plant species are advised on sloppy areas for soil and water conservation practices [37]. However, research in South Ethiopia, indicates that *Salix subserrata* is promising plant species for slope stabilization because it shows better root mechanical properties and has better root cohesion [38]. Nitrogen fixing ability of woody plants is the root chemical trait that centrally considered while integrating trees and shrubs in Agroforestry system to improve the soil nutrient [30, 39, 40]. Though tree species are used to improve soil fertility and combat soil erosion, there is limited knowledge regarding the root attribute of the woody plant species in the case of Ethiopia due to limited laboratory and well skilled manpower on the area [41, 42].

However, the belowground functioning of Agroforestry systems is still lacking, because numerous and complex site-specific interactions and trade-offs are at play [43]. For example, [44] state that the existence of *F. albida* tree highly improved N and P use efficiencies, leading to pointedly higher grain yields in wheat in the case of Mojo, the Central Rift Valley of Ethiopia. In contrast to this, *Gravilia robust* (*G. robusta*) (in Bugesera, Rwanda) and *Acacia tortilis* (*A. tortilis*) (in Meki, Ethiopia) trees lowered nutrient use efficiencies in maize, leading to significantly less maize grain yields compared with open fields receiving the same fertilization [44]. The research done on the effect of *F. albida* and *A. tortilis* on yield and biomass of wheat grown under canopies of both trees in Bora district Central Rift-valley, Ethiopia by [45] pointed out significant difference in nutrient availability between under canopy and open plot leading to greater grain yield under the canopy. The effect of the *F. albida (Del)* and *C. macrostachyus (Lam)* tree species on soil fertility parameters as well as grain yield of maize was significantly higher within the canopy of the tree than outside of the canopy in the case of umbulo Wacho watershed, southern Ethiopia [42]. Furthermore, [46] reported higher grain yield of sorghum (*Sorghum bicolor*) under canopy of *Faidherbia albida Delile* and *Cordia Africana Lam* trees species as compared to the open field in East Hararghe Zone, Oromia National Regional State; Ethiopia. Since these trees have the potential to improve soil fertility and moisture under its canopy. Ref. [47] also indicate yield of sorghum is significantly higher under the *F. albida* canopy than away from in the Tahtay Maychew district, central zone of Tigray region in the Northern part of Ethiopia.

The positive impact of trees on yield may be attributed by different factors. For example, there is higher Arbuscular Mycorrhizal Fungi under and at the periphery of the *F. albida* canopy than away from it and inhibit the growth and development of striga which is an obligate root hemi-parasitic weed of maize and sorghum [47]. Ref. [48] hold the same view when they state that an agroforestry system has increased abundance of soil bacteria and fungi and soil-N-cycling genes than monoculture cropland and open grassland. Other approach to minimize the negative effect of trees on crop yield is management. For instance, repeated tillage and weed management tended to minimize the negative impact of trees on crops, underlining the importance of agronomic practices that minimize competition between trees and crops for belowground resources [33].

Regarding to water use between trees and crops, [49] show that there is higher soil infiltrability under single trees than in the open areas indicating a positive impact of trees on soil hydraulic properties influencing groundwater recharge. Further, [50, 51] indicate the occurrence of plant hydrologic niche segregation in the agroforestry system suggesting weak competition for water between the components of the system. In coffee based Agroforestry system, [52] reported the coffee water uptake is mainly sustained from shallow soil sources (< 15 cm depth), while all shade trees relied on water sources from deeper soil layers (> 15 to 120 cm depth).

Concerning to allellopathic effect, different woody plant species produce different chemicals with allelopathic contents such as benzoic, cinnamic and phenolic acids, which have the potential to inhibit neighboring plants either positively or negatively depending on their concentrations [53]. The potential allelopathic effect of different Agroforestry tree species on Ethiopian main crops was studied in different parts of the country by different authors. For instance, the study conducted by [54] on the effects of four woody species on seed germination, radicle and seedling growth of four main Ethiopian crops namely; *Cicer arietinum* (chickpea), *Zea mays* (maize), *Pisum sativum* (pea) and *Eragrostis tef* (teff) reveals that Aqueous leaf extracts of all the tree species significantly reduced seedling growth, germination rate and radicle growth of the majority of the crops. Other studies conducted in the country also illustrate the effect of different woody species on different crops. For instance, chemical extracted from *Eucalyptus grandis* and *Eucalyptus camaldulensis* reduces germination rate, collar diameter, root length, and shoot length of Haricot Bean and Maize [55]. Ref. [56] reported also chemical extract from *Prosopis juliflora* reduces radicle and plumule length, *Z. mays*, *Panicum maximum*, *Chloris gayana*, *Gossypium hirsutum*. The adverse effect of chemicals from woody plant species on yield components like; shoot length, root length and collar diameter of crops significantly reduce the yield of the crops. In reverse to these studies, study conducted by [57] shows that, chemicals extracted from *Gravellia robusta* and *Casuarina equisetifolia* have stimulatory effect on the germination and radicle growth of Wheat and Maize.

3. Role of woody plants in soil fertility enhancement

Trees have impressive potential to improve soil fertility and forbid soil erosion in land management like farmland and watershed management [58–61]. For example, benefits of farmland woody plant species in the case of Northwestern Ethiopia are tremendous and soil fertility enhancement and management role indicates 35.14% among other benefits [62].

Dispersed trees on smallholder farms enhance soil fertility. For instance, research done in Tigray region reveals that *Dalbergia melanoxylon* woody species added a significant amount of nutrients to the soil indicating a negative linear relationship between the radial distance of the woody species and soil total nitrogen (TN), organic carbon (OC) and available phosphorus (AvP) contents (see **Table 1**) [60]. Ref. [42] investigated also a negative linear relationship between the radial distance of the woody species and soil TN, OC, and AvP contents for *F. albida* and *C. macrostachyus*. Ref. [64] shows that *Sesbania sesban* tree is also significantly (P < 0.05) improves soil TN, OC, Av.P, potassium (K), sodium (Na), calcium (Ca), and magnesium (Mg) of degraded lands from Lemo District, Hadiya Zone, Southern Nations, Nationalities and Peoples' Regional State (SNNPR) (see **Table 1**). Ref. [63] share a similar opinion when they state that higher K and Mg soil nutrients are observed under *Ficus vasta* and *Albizia gummifera* trees, compared with open fields, in parkland Agroforestry in central rift valley of Ethiopia (see **Table 1**).

Likewise, [66] states *Acacia abyssinica* specie is the most common in the crop-livestock farms in Borodo Watershed, Central Ethiopia because of its capacity to improve soil fertility and provide another service as well. According to the experience of farmers from highlands of the Kembatta zone, in the SNNPR, Ethiopia, Erythrina spp. and *Vernonia amygdalina* tree species are commonly grown in the hedges to improve soil fertility [67]. According to the Farmers' experience in Adola Rede District, Guji Zone, Southern Ethiopia, *Ficus sur* and *Cordia africana* tree species are found to be the most preferred coffee shade tree species for soil fertility improvement in the first and second rank respectively. Furthermore, the soil laboratory analyzed results indicates that soil chemical properties under canopy of both *Ficus sur* and *Cordia africana* shade tree species are in line with farmers' rank of shade tree preferences based on soil fertility improvement character (see **Table 2**) [65]. This is due to higher organic matter input through litter fall, root biomass, uptake and return of nutrients from deeper soil profiles under the tree canopies [68]. Furthermore, regulating services of trees on parklands are the protection of the soil against wind and water erosion, reduction of temperature through their shade as well as supporting services through improvement of soil fertility [69].

4. Role of woody plants in soil erosion control

Regarding to soil erosion control, biological soil and water conservation measures like tree and shrub planting are used to strengthen physical structures. The strengthened physical structure enabled to stabilize soil along the physical structures and to reduce the speed of surface runoff, henceforth increasing the infiltration rate of soil [70, 71]. Tree species that commonly being planted along soil and water conservation structures such as bunds and trenches namely are *Croton machrostachyus*, *Acacia abyssinica*, *Sesbania sesban* and *Vernonia amygdalina* [70]. *Sesbania sesban* species is the farmers' most preferred woody plant for planting in soil bund in the semi-arid sites in Oromia, Ethiopia while *Leucaena leucocephala* and *Sesbania sesban* are in the sub-humid sites of the same region [72]. Additionally, *Acacia abyssinica* (37%) and *Fiaderbia albida* (30%) were the top woody species that

Species name	Sample plots	Chemical properties of soil, exchangeable base (Meq/100 g soil)							References
		Na	K	Ca	Mg	OC%	TN%	AvP (ppm)	
Ficus vasta	CN	0.55 ± 0.21^a	2.64 ± 1.75^a	7.87 ± 1.84^a	2.81 ± 0.76^a	—	—	—	[63]
	OP	0.43 ± 0.23^a	1.47 ± 0.221^b	6.21 ± 1.48^a	2.44 ± 0.32^a	—	—	—	
Albizia gumifera	CN	0.69 ± 0.11^a	4.42 ± 1.65^a	12.41 ± 3.24^a	3.39 ± 1.76^a	—	—	—	
	OP	0.60 ± 0.28^a	1.86 ± 0.89^b	12.15 ± 2.45^a	3.27 ± 0.92^a	—	—	—	
Oxytenanthera abyssinica	CN	—	—	—	—	$1.73 (0.16)^a$	$0.26 (0.01)^a$	$7.21 (0.20)^a$	[60]
	NCN	—	—	—	—	$1.28 (0.09)^b$	$0.13 (0.01)^b$	$6.55 (0.19)^a$	
	FCN	—	—	—	—	$1.30 (0.11)^b$	$0.12 (0.01)^b$	$6.02 (0.21)^b$	
Dalbergia melanoxylon	CN	—	—	—	—	$1.02 (0.06)^a$	$0.13 (0.005)^a$	$6.37 (0.28)^a$	
	NCN	—	—	—	—	$0.70 (0.06)^b$	$0.09 (0.005)^b$	$5.78 (0.21)^{a-b}$	
	FCN	—	—	—	—	$0.65 (0.05)^b$	$0.07 (0.004)^c$	$5.32 (0.17)^b$	
Sesbania sesban	LS	0.05 ± 0.00^a	1.57 ± 0.10^a	25.48 ± 1.33^a	3.39 ± 0.17^a	2.37 ± 0.11^a	0.21 ± 0.01^a	3.85 ± 0.31^a	[64]
	LEG	0.042 ± 0.00^a	1.41 ± 0.18^a	24.14 ± 4.6^a	3.20 ± 0.6^a	2.17 ± 0.03^b	0.185 ± 0.0^b	$3.52 \pm 0.46^{a-b}$	
	DGL	0.032 ± 0.0^b	1.15 ± 0.18^b	17.58 ± 0.8^b	2.33 ± 0.11^b	1.97 ± 0.15^c	0.165 ± 0.0^c	2.86 ± 0.47^b	
Ficus sur	CN	—	$2.27^a \pm 0.95$	—	—	$6.49^a \pm 1.31$	$0.67^a \pm 0.15$	$7.52^a \pm 1.87$	[65]
	OP	—	$0.41^b \pm 0.32$	—	—	$2.54^b \pm 0.65$	$0.41^b \pm 0.12$	$3.81^b \pm 0.91$	
Cordia africana	CN	—	$1.05^a \pm 1.15$	—	—	$4.51^a \pm 1.15$	$0.49^a \pm 0.09$	$4.58^a \pm 0.85$	
	OP	—	$0.56^b \pm 0.24$	—	—	$2.31^b \pm 0.91$	$0.42^b \pm 0.07$	$2.50^b \pm 0.41$	

Species name	Sample plots	Chemical properties of soil, exchangeable base (Meq/100 g soil)							References
		Na	K	Ca	Mg	OC%	TN%	AvP (ppm)	
F. albida	1.5 m distance from the canopy	0.34 (0.03)[a]	1.33 (0.32)[a]	42.05 (1.83)[a]	13.22 (2.29)[a]	2.03 (0.21)[a]	0.41 (0.03)[a]	11.33 (0.6)[a]	[42]
	3.5 m distance from the canopy	0.27 (0.09)[a,b]	1.13 (0.3)[a]	39.04 (1.7)[a]	11.21 (2.1)[a]	1.49 (0.32)[b]	0.31 (0.04)[b]	10.03 (0.4)[a]	
	25 m distance from the canopy	0.24 (0.06)[b]	0.79 (0.16)[b]	29.38 (0.79)[b]	8.70 (0.66)[b]	1.38 (0.29)[b]	0.23 (0.03)[c]	8.73 (0.47)[b]	
Croton machrostachyus	1.5 m distance from the canopy	0.31 (0.04)[a]	0.90 (0.14)[a]	36.94 (8.31)[a]	10.25 (1.12)[a]	1.26 (0.25)[a]	0.14 (0.00)[a]	9.03 (1.08)[a]	
	3.5 m distance from the canopy	0.28 (0.07)[a,b]	0.84 (0.22)[a]	34.16 (8.8)[a]	10.81 (0.82)[a]	1.03 (0.16)[b]	0.13 (0.04)[b]	8.71 (0.74)[b]	
	25 m distance from the canopy	0.26 (0.04)[b]	0.49 (0.11)[b]	24.64 (3.54)	9.88 (0.45)[b]	0.76 (0.09)[b]	0.08 (0.01)[c]	8.47 (0.55)[b]	

[a,b,c] Means followed by different letters are significantly different.
CN, under woody species canopy; OP, open field; NCN, near to canopy; FCN, far from canopy; LS, lands treated with Sesbania; LEG, lands treated with elephant grass; DGL, degraded grazing land.

Table 1.
Impact of woody plant species on chemical properties of soil in the case of Ethiopia.

Species	Sample plots	Chemical properties of soil, Exchangeable base (Meq/100 g soil)							References
		Na	K	Ca	Mg	OC%	TN%	AvP (ppm)	
Ficus sur	CN	—	2.27[a] ± 0.95	—	—	6.49[a] ± 1.31	0.67[a] ± 0.15	7.52[a] ± 1.87	[65]
Cordia africana	CN	—	1.05[b] ± 1.15	—	—	4.51[b] ± 1.15	0.49[b] ± 0.09	4.58[b] ± 0.85	

[a,b]*Means followed by different letters are significantly different.*

Table 2.
Comparison of the impact of woody plant species on chemical properties of soil in the case of Ethiopia.

practiced by farmers for soil conservation purpose in West Hararghe Zone, Oromia National Region State, Ethiopia [27].

5. Conclusions and recommendations

Woody plants of Agroforestry system improve soil fertility and forbid soil erosion from farmlands/water shade. Therefore, integration of woody plants on farming system based on the functional trait of woody plant is crucial to sustain soil management benefits of woody plants in Agroforestry systems.

Based on this review, the following are recommended to researchers to undertake study and policy makers to design agroforestry system that enable farmers to fully utilize the woody plant species potential in the Agroforestry system from functional trait point of view to achieve sustainable soil management practice in Ethiopia.

5.1 For researchers

1. Woody plant phenology such as leaf fall and re-growing and flowering seasons should be clearly investigated as per the Agro ecology because tree phenology is differing per Agro-ecology of the country. There is also lack of clear data on the phenology of major agroforestry woody plant species.

2. Woody plants' litter decomposability and their chemical compositions should be investigated further. Similarly, [73] recommend the importance of woody plants' litter decomposability and their chemical compositions analysis because litter quality is one among various factors which affects soil fertility based on its type and chemical contents.

3. The root system of woody plants used for soil and water conservation practice should be investigated.

4. The significance of the use of single species versus multiple species for soil nutrient improvement and soil erosion control should be evaluated.

5. Tree management practices of Parkland Agroforestry system to increase crop yield.

5.2 For policy makers

Woody plant functional traits should be considered when policy is designed to ensure sustainable soil management benefits of woody plants while introducing Agroforestry technologies.

Acknowledgements

We would like to express our special gratitude and thanks to Mr. Melkamu Teklu Kisi for his constructive comments and guidance during this work. Our gratitude and thanks also goes to Darko Hrvojic who invites and remind us to send our work to new book project "Biodiversity of Ecosystems" an Open Access book edited by Dr. Levente Hufnagel.

Conflict of interest

The authors declare that they have no known competing financial interests or personal relationships that could have appeared to influence the work reported in this paper.

Author details

Hana Tamrat Gebirehiwot*, Alemayehu Abera Kedanu and Megersa Tafesse Adugna
Department of Natural Resource Management, College of Agriculture and Natural Resource, Salale University, Fiche, Ethiopia

*Address all correspondence to: hanatamrat87@gmail.com

IntechOpen

References

[1] Violle C., Navas M.L., Vile D., Kazakou E., Fortunel C, Hummel I., Garnier E. (2007). Let the concept of trait be functional. Oikos 116: 882-892.

[2] Poorter, L. et al. (2015) 'Effects on competition', Nature, pp. 1-15. DOI:10.1038/nature16476.

[3] Mbow, C. et al. (2014) 'Science direct agroforestry solutions to address food security and climate change challenges in Africa', Current Opinion in Environmental Sustainability, 6, pp. 61-67. DOI:10.1016/j.cosust.2013. 10.014.

[4] Leakey, R. R. B. (2017) 'Definition of Agroforestry Revisited', in Multifunctional Agriculture, pp. 5-6. DOI:10.1016/b978-0-12-805356-0. 00001-5.

[5] Hailemariam, S. N., Soromessa, T. and Teketay, D. (2016) 'Land use and land cover change in the bale mountain eco-region of Ethiopia during 1985 to 2015', Land, 5(4). DOI:10.3390/land5040041.

[6] Miheretu, B. A. and Yimer, A. A. (2018) 'Land use/land cover changes and their environmental implications in the Gelana sub-watershed of northern highlands of Ethiopia', Environmental Systems Research, 6(1). DOI:10.1186/s40068-017-0084-7.

[7] Hundera, H., Mpandeli, S. and Bantider, A. (2020) 'Spatiotemporal analysis of land-use and land-cover dynamics of Adama District, Ethiopia and its implication to greenhouse gas emissions', Integrated Environmental Assessment and Management, 16(1), pp. 90-102. DOI:10.1002/ieam.4188.

[8] Tokuma, U. and Debissa, L. (2020) 'Spatiotemporal Landuse Land Cover Changes in Walmara', 9(1), pp. 32-37. DOI:10.11648/j.earth.20200901.14.

[9] Gashaw, T. et al. (2017) 'Evaluation and prediction of land use/land cover changes in the Andassa watershed, Blue Nile Basin, Ethiopia', environmental Systems Research, 6(1). DOI:10.1186/s40068-017-0094-5.

[10] Deribew, K. T. and Dalacho, D. W. (2019) 'Land use and forest cover dynamics in the North-Eastern Addis Ababa, central highlands of Ethiopia', Environmental Systems Research, 8(1), pp. 1-18. DOI:10.1186/s40068-019-0137-1.

[11] Gebremedhin, Y. G. (2019) 'Soil Erosion Hazard in Errer Dembel Sub-Basin, in Shinille zone of the Ethiopia Somali regional state', international journal of environmental sciences and natural Resources, 17(1), pp. 1-9. DOI:10.19080/ijesnr.2019.17. 555951.

[12] Bishaw, B. and Abdelkadir, A. (2003) 'Agroforestry and Community Forestry for Rehabilitation of Degraded Watersheds on the Ethiopian Highlands', Combating Famine in Ethiopia (October), pp. 1-22.

[13] Martin, A. R. and Isaac, M. E. (2015) 'Plant functional traits in agroecosystems: A blueprint for research', Journal of Applied Ecology, 52(6), pp. 1425-1435. DOI:10.1111/1365-2664.12526.

[14] Pérez-Harguindeguy, N. et al. (2013) 'New handbook for standardised measurement of plant functional traits worldwide', Australian Journal of Botany, 61(3), pp. 167-234. DOI:10. 1071/BT12225.

[15] Haggerty, B. P. and Mazer, S. J. (2008) 'The Phenology Handbook'.

[16] Rodriguez, H. G., Maiti, R. and Sarkar, N. C. (2014) 'Phenology of Woody species : A review phenology of

Woody species: A review' (January). DOI:10.5958/0976-4038.2014.00595.8.

[17] Kassam, A. et al. (2013) 'Crops are grown', pp. 337-400.

[18] Guyassa, E. and Raj, A. J. (2013) 'Assessment of biodiversity in cropland agroforestry and its role in livelihood development in dryland areas: A case study from Tigray region, Ethiopia', Journal of Agricultural Technology, 9(4), pp. 829-844.

[19] Molla, A. and Kewessa, G. (2015) 'Woody Species Diversity in Traditional Agroforestry Practices of Dellomenna District, Southeastern Ethiopia: Implication for Maintaining Native Woody Species', International Journal of Biodiversity, 2015(iii), pp. 1-13. DOI:10.1155/2015/643031.

[20] Buchura, N. W., Debela, H. F. and Zerihun, K. (2019) 'Assessment of woody species in agroforestry systems around Jimma town, Southwestern Ethiopia', International Journal of Biodiversity and Conservation, 11(1), pp. 18-30. DOI:10.5897/ijbc2018.1207.

[21] Yasin, H., Kebebew, Z. and Hundera, K. (2018) 'Woody species diversity, regeneration and socioeconomic benefits under natural forest and adjacent coffee agroforests at belete Forest, Southwest Ethiopia', Ekologia Bratislava, 37(4), pp. 380-391. DOI:10.2478/eko-2018-0029.

[22] Kassa, H. et al. (2018) 'Agro-ecological implications of forest and agroforestry systems conversion to cereal-based farming systems in the White Nile Basin, Ethiopia', Agroecology and Sustainable Food Systems. Taylor and Francis, 42(2), pp. 149-168. DOI:10.1080/21683565.201 7.1382425.

[23] Gebru, B. M. et al. (2019) 'Socio-ecological niche and factors affecting agroforestry practice adoption in different agroecologies of southern Tigray, Ethiopia', Sustainability (Switzerland), 11(13), pp. 1-19. DOI:10.3390/su11133729.

[24] Young, A. (1990) 'Maintenance of soil fertility for sustainable production of trees and crops through Agrof ores try systems sustainability and soil conservation', Japan International Research Center for Agricultural Sciences, pp. 198-206.

[25] Linger, E. (2014) 'Agro-ecosystem and socio-economic role of homegarden agroforestry in Jabithenan District, North-Western Ethiopia: Implication for climate change adaptation', SpringerPlus, 3(1), pp. 1-9. DOI:10.1186/2193-1801-3-154.

[26] Bussa, B. and Feleke, K. (2020) 'Contribution of Parkland Agroforestry Practices to the Rural Community Livelihood and Its Management in', 8(4), pp. 104-111. DOI:10.11648/j.hss. 20200804.11.

[27] Yusuf, H. and Solomon, T. (2019) 'Woody Plant Inventory and Its Management Practices in Traditional Agroforestry of West Hararghe Zone, Oromia National Region State, Ethiopia', 8(5), pp. 94-103. DOI:10.11648/j. ajep.20190805.11.

[28] Ernstberger, J., 2017. Perceived Multifunctionality of Agroforestry Trees in Northern Ethiopia.

[29] Ssebulime, G. et al. (2018) 'Canopy management, leaf fall and litter quality of dominant tree species in the banana agroforestry system in Uganda', African Journal of Food, Agriculture, Nutrition and Development, 18(1), pp. 13154-13170. DOI:10.18697/ajfand.81.16700.

[30] Hundera, K. (2016) 'Shade tree selection and management practices by farmers in traditional coffee production systems in Jimma Zone, Southwest Ethiopia', Ethiopian Journal of

Education and Sciences, 11(2), pp. 91-105-105.

[31] Muleta, D. et al. (2011) 'Organic benefits of shade trees un coffee production systems in Bonga and Yayu-Hurumi districtis southwestern Ethiopia: Farmers' perception', Ethio. J. Educ. And Sc (1).

[32] Zhao, B. et al. (2019) 'Effects of Rainfall Intensity and Vegetation Cover on Erosion Characteristics of a Soil Containing Rock Fragments Slope', Advances in Civil Engineering, 2019. DOI:10.1155/2019/7043428.

[33] Sida, T. S., Baudron, F., Hadgu, K., Derero, A., and Giller, K. E. (2018). Crop vs. tree: Can agronomic management reduce trade-offs in tree-crop interactions? Agriculture, Ecosystems and Environment, 260(July 2017), 36-46.

[34] Haile, G., Lemenih, M., Itanna, F., and Agegnehu, G. (2021). Comparative study on the effects of Acacia albida on yield and yield components of different cereal crops in southern Ethiopia. Acta Agriculturae Scandinavica, Section B—Soil and Plant Science, 1-13.

[35] Ndoli, A., Baudron, F., Sida, T. S., Schut, A. G. T., van Heerwaarden, J., and Giller, K. E. (2018). Conservation agriculture with trees amplifies negative effects of reduced tillage on maize performance in East Africa. Field Crops Research, 221(March), 238-244. DOI:10.1016/j.fcr.2018.03.003

[36] Kuria, A. et al. (2013) 'Local Knowledge of Farmers on Opportunities and Constraints to Sustainable Intensification of Crop – Livestock – Trees Mixed Systems in Basona Woreda, Amhara Region, Ethiopian Highlands', p. 72.

[37] Ghestem, M. et al. (2014) 'A framework for identifying plant species to be used as "ecological engineers" for

fixing soil on unstable slopes', PLoS ONE, 9(8). DOI:10.1371/journal.pone.0095876.

[38] Tsige, D., Senadheera, S. and Talema, A. (2020) 'Stability analysis of plant-root-reinforced shallow slopes along mountainous road corridors based on numerical modeling', Geosciences (Switzerland), 10(1), pp. 25-37. DOI:10.3390/geosciences10010019.

[39] Sharma, K. L. (2008) 'Effect of agroforestry systems on soil quality – Monitoring and assessment', academia. Edu, pp. 122-132.

[40] Mehari, A. (2012) 'Traditional agroforestry practices, opportunities, threats and research needs in the highlands of Oromia, Central Ethiopia', International Research Journal of Agricultural Science and Soil Science, 2(5), pp. 194-206.

[41] Reubens, B. et al. (2011) 'Tree species selection for land rehabilitation in Ethiopia: From fragmented knowledge to an integrated multi-criteria decision approach', Agroforestry Systems, 82(3), pp. 303-330. DOI:10.1007/s10457-011-9381-8.

[42] Manjur, B., Abebe, T. and Abdulkadir, A. (2014) 'Effects of scattered F. albida (Del) and C. macrostachyus (lam) tree species on key soil physicochemical properties and grain yield of maize (Zea Mays): A case study at umbulo Wacho watershed, southern Ethiopia', Wudpeckers J. Agric. Resear., 3(3), pp. 63-73.

[43] Cardinael, R., Mao, Z., Chenu, C., and Hinsinger, P. (2020). Belowground functioning of agroforestry systems: Recent advances and perspectives. Plant andSoil,453(1-2),1-13.DOI:10.1007/s11104-020-

[44] Sida, T. S., Baudron, F., Ndoli, A., Tirfessa, D., and Giller, K. E. (2020). Should fertilizer recommendations be adapted to parkland agroforestry

systems? Case studies from Ethiopia and Rwanda. Plant and Soil, 453(1), 173-188.

[45] Desta, K. N. (2018). Wheat yields under the canopies of Faidherbiaalbida (Delile) a. Chev and Acacia tortilis (Forssk.) Hayenin Park land agroforestry system in central Rift Valley, Ethiopia. Agriculture, forestry and Fisheries, 7(3), 75.

[46] Abdella, M., Nigatu, L., and Akuma, A. (2020). Impact of parkland trees (Faidherbia albida Delile and Cordia Africana lam) on selected soil properties and Sorghum yield in eastern Oromia, Ethiopia. Agriculture, Forestry and Fisheries, 9(3), 54.

[47] Birhane, E., Gebremeskel, K., Taddesse, T., Hailemariam, M., Hadgu, K. M., Norgrove, L., and Negussie, A. (2018). Integrating Faidherbia albida trees into a sorghum field reduces striga infestation and improves mycorrhiza spore density and colonization. Agroforestry Systems, 92(3), 643-653.

[48] Beule, L., Corre, M. D., Schmidt, M., Göbel, L., Veldkamp, E., and Karlovsky, P. (2019). Conversion of monoculture cropland and open grassland to agroforestry alters the abundance of soil bacteria, fungi and soil-N-cycling genes. PloS one, 14(6), e0218779.

[49] Bargués Tobella, A., Reese, H., Almaw, A., Bayala, J., Malmer, A., Laudon, H., and Ilstedt, U. (2014). The effect of trees on preferential flow and soil infiltrability in an agroforestry parkland in semiarid Burkina Faso. Water resources research, 50(4), 3342-3354.

[50] Wu, J., Liu, W., and Chen, C. (2016). Below-ground interspecific competition for water in a rubber agroforestry system may enhance water utilization in plants. Scientific reports, 6(1), 1-13.

[51] Wu, J., Zeng, H., Chen, C., Liu, W., and Jiang, X. (2019). Intercropping the sharp-leaf galangal with the rubber tree exhibits weak belowground competition. Forests, 10(10), 924.

[52] Muñoz-Villers, L. E., Geris, J., Alvarado-Barrientos, M. S., Holwerda, F., and Dawson, T. (2020). Coffee and shade trees show complementary use of soil water in a traditional agroforestry ecosystem. Hydrology and Earth System Sciences, 24(4), 1649-1668.

[53] Iqbal, J., Rauf, H. A., Shah, A. N., Shahzad, B., and Bukhari, M. A. (2017). Allelopathic effects of rose wood, guava, eucalyptus, sacred fig and jaman leaf litter on growth and yield of wheat (*Triticum aestivum* L.) in a wheat-based agroforestry system. Planta Daninha, 35.

[54] Nigatu, L., and Michelsen, A. (1992). Allelopathy in agroforestry systems: The effects of leaf extracts of Cupressus lusitanica and three Eucalyptus species. on four Ethiopian crops. Agroforestry Systems, 21, 63-74.

[55] Gurmu, W. R. (2015). Effects of aqueous Eucalyptus extracts on seed germination and seedling growth of Phaseolus vulgaris L. and Zea mays L. open access Library Journal, 2(09), 1.

[56] Asrat, G., and Seid, A. (2017). Allelopathic effect of Meskit (Prosopis juliflora (Sw.) DC) aqueous extracts on tropical crops tested under laboratory conditions. Momona Ethiopian Journal of Science, 9(1), 32-42.

[57] Ayalew, A., and Asfaw, Z. Allelophatic effects of Gravellia Robusta, Eucalyptus Camaldulensis and *Casuarina equisetifolia* on Germination and Root Length of Maize and Wheat.

[58] Mushir, A. and Kedru, S. (2012) 'Soil and water conservation management through indigenous and traditional practices in Ethiopia: A case study', Ethiopian Journal of

Environmental Studies and Management, 5(4). DOI:10.4314/ejesm.v5i4.3.

[59] Belayneh, M., Yirgu, T. and Tsegaye, D. (2019) 'Effects of soil and water conservation practices on soil physicochemical properties in Gumara watershed, upper Blue Nile Basin, Ethiopia', Ecological Processes. Ecological Processes, 8(1). DOI:10.1186/s13717-019-0188-2.

[60] Gebrewahid, Y. et al. (2019) 'Dispersed trees on smallholder farms enhance soil fertility in semi-arid Ethiopia', Ecological Processes. Ecological Processes, 8(1). DOI:10.1186/s13717-019-0190-8.

[61] Kuyah, S. et al. (2019) 'Agroforestry delivers a win-win solution for ecosystem services in sub-Saharan Africa. A meta-analysis', agronomy for sustainable development. Agronomy for Sustainable Development, 39(5). DOI:10.1007/s13593-019-0589-8.

[62] Giday, K. et al. (2019) 'Studies on farmland woody species diversity and their socioeconomic importance in Northwestern Ethiopia', Tropical Plant Research, 6(2), pp. 241-249. DOI:10.22271/tpr.2019.v6.i2.34.

[63] Asfaw, Z. et al. (2016) 'Development Research Woody Species Composition and Soil Properties Under Some Selected Trees in Parkland Agroforestry in Central Rift Valley of Ethiopia'.

[64] Sinore, T., Kissi, E. and Aticho, A. (2018) 'International Soil and Water Conservation Research The effects of biological soil conservation practices and community perception toward these practices in the Lemo District of Southern', International Soil and Water Conservation Research. Elsevier B.V., 6(2), pp. 123-130. DOI:10.1016/j.iswcr.2018.01.004.

[65] Emire, A. (2018) 'Status of soil properties under canopy of farmers' preferred coffee shade tree species, in Adola Rede District, Guji zone, southern Ethiopia', American Journal of Agriculture and Forestry, 6(5), p. 148. DOI:10.11648/j.ajaf.20180605.15.

[66] Sisay, M. (2013) 'Tree and shrub species integration in the crop-livestock farming system', Tree and Shrub Species Integration in the Crop-Livestock Farming System, 21(1), pp. 647-656. DOI:10.4314/acsj.v21i1.

[67] Astrid, M. (2019) Internship Thesis GEEFT 2018-2019, Hedgerows and Agroforestry Practices in the Highlands of Kembatta zone, Ethiopia

[68] Wolle, H. S., Lemma, B., and Mengistu, T. (2019). Effects of Ziziphus spina-christi (L.) on selected soil properties and sorghum yield in Habru District, north Wollo, Ethiopia. Malaysian Journal of Medical and Biological Research, 6(2), 85-92.

[69] Sanogo, K., Binam, J., Bayala, J., Villamor, G. B., Kalinganire, A., and Dodiomon, S. (2017). Farmers' perceptions of climate change impacts on ecosystem services delivery of parklands in southern Mali. Agroforestry systems, 91(2), 345-361.

[70] Kuria, A. et al. (2014) 'Local Knowledge of Farmers on Opportunities and Constraints to Sustainable Intensification of Crop-Livestock-Trees Mixed Systems in Lemo Woreda, SNNPR Region, Ethiopian Highlands'.

[71] Gemechu, T. and Hunde, K. K. (2015) 'Assessment on Farmers' Practices on Soil Erosion Control and Soil Fertility Improvement in Rift Valley Areas of East Shoa and West Arsi Zones of Oromia, Ethiopia''', EC Agriculture 2.4, 4(March), pp. 391-400.

[72] Derero, A. et al. (2020) 'Farmer-led approaches to increasing tree diversity

in fields and farmed landscapes in
Ethiopia', Agroforestry Systems, 7.
DOI:10.1007/s10457-020-00520-7.

[73] Mohammed, M., Beyene, A. and
Reshad, M. (2018) 'Influence of
Scattered Cordiaafricana and
Crotonmacrostachyus Trees on Selected
Soil Properties, Microclimate and Maize
Yield in Eastern Oromia, Ethiopia',
American Journal of Agriculture and
Forestry. Vol. 6, No. 6, 2018, pp. 253-262.
DOI:10.11648/j.ajaf.20180606.23, 6(6),
pp. 253-262.

Chapter 10

Soil Biodiversity as a Key Sponsor of Regenerative Agriculture

Mulugeta Aytenew

Abstract

Increasing knowledge and literacy around soil biodiversity is essential to discover and implement biological solutions for the discouraging challenges people face in agriculture and human wellbeing. Therefore, this review was done to get an insight into the awareness and understanding of the contribution of soil biodiversity to regenerative agriculture. The review was done by referring to the latest different research findings; reports, working guidelines, as well as knowledge shared from different soil biodiversity conferences and webinar discussion points. The review disclosed that to meet the increasing demand for food for the ever-increasing global population and the 2030 sustainable development goals, regenerating the already degraded lands through regenerative agriculture principles and practices is vitally important. The findings and report documents showed that soil biodiversity facilitates the regenerative agriculture system as soil organisms are using as soil health improvement machines, a remediates for soil and water pollution, a fertilizer, pesticide, as a means of carbon sink, and used in the pharmaceutical industry to discover new drugs and vaccines for animal and human health. Moreover, the meta-analysis publicized that the consideration and use of soil biodiversity in the regenerative agriculture system have promising results although little is known about the role of those soil organisms in the ecosystem due to the presence of knowledge gap and complexity of relationships in the soil system. Therefore, furthermore, attention should be given to the discoveries of soil biodiversity to use them as a natured based solution for regenerative agriculture in the 21st century and to meet the 2030 sustainable development goals.

Keywords: Bio-fertilizer, Soil biodiversity, Ethiopia, Holistic systems approach, Nature-based solution

1. Introduction

Agriculture is a soil-based industry that is heavily burdened to feed the increasing global population. And soil has been described as "the fragile living skin of the Earth", but yet both its aliveness and fragility have often been ignored in the expansion of agriculture across the face of the globe [1]. Since it is a pivotal component in a global nexus of soil, water, air, and energy, how we treat the soil can impact massively on agriculture and climate change. Soils constitute one of the largest reservoirs of biodiversity on Earth and soil organisms are the source of key ecological functions and services that support agriculture, including soil conservation, water cycling, pest and disease regulation, carbon sequestration, and nitrogen fixation [2].

Currently, sustainable agriculture which is expressed as "industrialized agriculture" [3] relies on monoculture cropping, increasing use of mechanization, the application of synthetic fertilizers, pesticides, and herbicides, along with liberal government subsidies. Although this approach can be considered successful, in that it has managed to feed a massively rising human population [4], a range of environmental and social burdens have also been incurred, including erosion, soil nutrient depletion and contamination; loss of water resources and biodiversity, loss of forests and desertification; human labor abuses; and naturally the decline of the traditional farming practices.

In Ethiopia, one of the most singly destructive factors in farming is land resource degradation [5]. Almost all the land resource balances in Ethiopia show a soil nutrient deficit, water and soils are eroded, forests are depleted, wildlife and biodiversities are disturbed; representing a loss of yield and quality for consumption and causes climate change. Once the land becomes degraded, fertility and health of the soil are lost; farmers suffer extreme losses in very low yields on their farms. Such losses are projected in an environment sensitive to climate change, cost of living, and starvation. Hence, urgent steps are needed to avoid this and regenerate the depleted resources.

Therefore, there is no need of sustaining the already degraded land resources, rather regenerating them and formulating sustainable agriculture. The agricultural revolution in Ethiopia is wishing for effective solutions which are fundamental to land management and agricultural practices. Regenerative Agriculture which defined as "*a holistic systems approach to agriculture that encourages continual on-farm innovation for environmental, social and economic wellbeing and improves the land resources it uses, rather than destroying or depleting them* [6, 7] is crucial for successful land management and agricultural practices.

Regenerative agriculture at its core has the intention to improve the health of soil or to restore highly degraded soil, which symbiotically enhances the quality of water, vegetation, and land productivity [8, 9]. Essentially, regenerative agriculture depends upon soil biodiversity and there may be no soil biodiversity without practices of regenerative agriculture; they have evolved together. By using methods of regenerative agriculture, it is possible not only to increase the amount of soil organic carbon in existing soils but to build new soils through attenuate the rate of soil erosion, restoration of the soil food web, improvement of soil fertility, and the activities of plants, animals, insects, fungi, bacteria, and humans too, all play a part in the formation of soil [1]. Hence, for the future scenarios challenging the agricultural sector such as soil degradation, increasing food demand, climate change, water scarcity, global soil biodiversity education and consideration of soil biology as a long term solution is needed to realize the full benefits of regenerative agriculture and respond to the needs of farmers and consumers relating to agriculture and land management.

Similarly, soil biodiversity plays a role in the formation of soil and enhances the ecosystem functions, services [10], and intern production and productivity of regenerative agriculture. Thus, this would lead to the consideration of soil biodiversity (activities of plant roots, animals, insects, fungi, bacteria ...) as nature-based solutions in the restoration of the soil food web, improvement of soil health and fertility, agricultural productivity, while locking-up carbon from the atmosphere.

Increased education and awareness are key strategies in ensuring that soil biodiversity is no longer out of sight, out of mind. As agricultural soils are under threat, there is a need to promote interactions between scientists, policymakers, and the general public to transfer and implement scientific findings of the benefits of soil biodiversity and ways to restore and conserve it [11]. These organizations and Elizabeth *et al.*, [12] also asserted that soil biodiversity is critical for soil functioning and plant production but has been largely ignored in global, regional, and national policies that address land management, food security, climate change, loss of bio-diversity, and desertification.

And finally, increasing knowledge and literacy, and passion, particularly around soil biodiversity is essential to draw on the diverse community of stakeholders required to discover and implement biological solutions for the daunting challenges people face in climate change, agriculture, ecosystem restoration, environmental pollution, and human health. Once more, this review is the place to get awareness and understanding of the contribution of soil biodiversity to regenerative agriculture.

2. Methodology

The review was done by a literature search and document sourcing using an online search in major websites that provide access to scientific research, like Research Gate, Science Direct, and Google Scholar to referring different research findings; reports, and working guidelines, as well as knowledge shared from differ-ent soil biodiversity conferences and webinar discussion points. Besides, citations in key papers were followed to identify additional relevant Articles and synthesize relevant peer-reviewed articles and related literature. Hence, it may represent a general diversity of regions and nations and provides a wealth of principles, exam-ples, actions, and solutions to bring soil biodiversity into the light of regenerative agriculture.

3. Contribution of soil biodiversity for regenerative agriculture

3.1 Regenerative agriculture: overview

Current conventional farming methods are resulting in the loss of fertile soil and biodiversity. According to Maria-Helena Semedo of the FAO, as cited by Chris [13], the world could run out of topsoil in about 60 years if we continue at current rates of soil destruction, as now about a third of the world's soil has already been degraded. This affects the earth's ability of food production, water filtering, carbon absorption, and farmers will no longer have enough arable topsoil to feed the growing world population. There might be a duty to transit towards regenerative agricultural practices.

Regenerative Agriculture is a system of farming principles and practices that increases biodiversity, enriches soils, improves watersheds, and enhances ecosys-tem services. The regenerative farming approach focuses on restoring soils that have been degraded by the industrial agricultural system. Its methods promote healthier ecosystems by rebuilding soil organic matter through holistic farming and grazing techniques. It enables the regeneration of land resources through the resto-ration of vegetation in a farm landscape using a high diversity of both annual and perennial crops [14, 15]. Moreover, it considers potential environmental and social impacts by eliminating the use of synthetic inputs and replacing them with site-specific management practices that maintain and increase long-term soil health, employment opportunities, and mitigation and adaptation to climate change.

André Leu [16] *who is the international director of Regeneration International* claims that transitioning 10–20% of agricultural production to the best practice of regen-erative systems (Biologically Enhanced Agricultural Management) will sequester enough carbon dioxide (2.3 ppm of CO_2 per year) to reverse climate change and can change agriculture from being a major contributor to climate change to becoming a major solution. There is broad agreement that most regenerative agriculture practices are good for soil health and have other environmental benefits [17, 18].

Among the regenerative agriculture principles, No-till reduces soil erosion and encourages soil water infiltration [19]. Cover crops do the same, and can also reduce water pollution and contribute to reducing soil organic matter losses [20]. Diverse crop rotations can lower pesticide use [21] and reduce environmental pollution [22]. Focuses strongly on the environmental dimension of sustainability, which includes themes such as *enhance and improve soil health, optimize resource management, alleviate climate change, improve nutrient cycling* and *water quality and availability,* articulated by improving soil health through soil biological activities [23].

In the experimental research of La Canne and Lundgren [24], regenerative corn fields generate nearly twice the profit of conventionally managed cornfields. Similarly, their finding discloses the insecticide-treated cornfields had higher pest abundance than untreated, regenerative cornfields. Reports from Burgess *et al.* [25]; IPCC [26] and Lunn-Rockliffe *et al.* [27] have stated the fundamental importance of transitioning to more regenerative agriculture methods if the world needs to meet its climate change targets, food security demands, protect farmland and build a healthier food system.

For the goal of agricultural development in 3rd world countries, the future agricultural production systems should be designed to take better advantage of production resources found on the farm [28, 29]. While most of the regenerative and organic markets are in developed countries, developing countries like Ethiopia are becoming important suppliers, as regenerative agriculture practices are particularly suited for the conditions of their farmers, especially smallholders living in rainfed areas. However, yet Government agencies in developing countries cannot often make the corporate sector responsible for agricultural development and for preventing harm to the environment. According to reports made by EPAT [30], pesticides that are illegal in Europe are commonly applied throughout sub-Saharan Africa, owing to the industry's open-door pesticide policy. Farmers in resource-constrained and low potential areas of Ethiopia traditionally use few external inputs [31, 32] but many of the environmental, social, and economic benefits of land management, which translate into ecological intensification, are hampered by a lack of appropriate regenerative agriculture knowledge and skills.

Therefore, by understanding and implementing regenerative agriculture; considering soil biodiversity; the farming community will benefit from enhanced nitrogen fixation, greater total organic matter production, nature-based pest management, genetic tolerance to stress conditions, and higher levels of biological activity all contribute to resource use efficiency and quality of products.

3.2 Importance of soil biodiversity as a nature-based solution

The sustainable development goal (SDG) which were adopted by the United Nations in 2015 as a universal call to action to end poverty, protect the planet, and ensure that by 2030 all people enjoy peace and prosperity provides a renewed motivation for focusing on using soil biodiversity for food and nutrition, and for linking it to the sustainability of future agricultural systems [22]. Soil biodiversity is critical for human health, plant growth and support, water and climate regulation, and erosion and disease control so as considered to be a common ground for achieving sustainability goals [12]. Hence, management and conservation of life in the soil are integral to governmental actions to provide healthy food, reduce greenhouse gases, lessen desertification and soil erosion, and prevent disease thereby regenerating agriculture.

According to the FAO and ITPS's Status of the World's Soil Resources report (2015) [33], soil organic carbon and soil biodiversity are crucial to increase food availability and the soil's ability to buffer against climate change effects. On the International Day for Soil biologic diversity (May 22, 2020), Semedo, who's the Deputy Director-General for primary natural resources of FAO highlighted *"there is a need to change the way to connect with nature, and that we need healthy soils for a healthy planet with vibrant ecosystems that allow our food system to be more resilient"* [34]. And these invisible organisms react, they play a crucial role in sustaining our planet, provide our food, supporting our health, our ways of living, and also human wellbeing. The issues of soil biodiversity become under the sort of focus of FAO [11] and of course the science that the potential role of soil is not just for food production and food productivity, but more important is about the environmental services and about the health of the planet and how the microorganisms in the soil play a very critical role in regeneration system of agriculture.

However, unfortunately, most of the case has not been explored so far, the knowledge that has about the soil biodiversity and soil mechanism biodiversity is really near nothing compared to the whole complexity that we have in all parts of the board and the sort of ecosystem. The United Nations in 2015 declared the year to the interest as the International Year of Soils and has asked FAO and the Global Soil Partnership to carry out the first global soil biodiversity assessment which is now in progress.

In agriculture, we have high productivity on the open networks soil. So, soils with biodiversity open networks have more productivity than soil with closed networks. So in nature, plants that are growing in these open networking sites (high diversity soil systems) are capable of taking out nutrients in an efficient way. Of course, the point is that we should not only increase soil biodiversity but also that we have to talk to crop breeders and agronomists to get the right crop species and crop varieties to grow on these biodiverse soils.

On the webinar held among 1136 participants on May 22, 2020, by representing more than 140 countries, around 72% of the people said that soil biodiversity is applied especially in crop production in their country. Then some have in ecosystem restoration, pollution and bioremediation, food processing, and very few in terms of the medical sector [34].

Going forward, harnessing natural resources (microbes, fauna, flora) together with SOM, is considered as the most effective approach for a sustainable increase in farm productivity, mitigating climate change, and restoring degraded environments [11]. Further evidence of the relationships between soil biodiversity and functioning concerning soil organic carbon (SOC) dynamics and primary productivity at farm scales can help in bridging the knowledge gaps in the biotic regulation of SOC turnover and plant productivity. This will represent a major advancement, not only in ecology but also in agriculture in the context of global climate change and food security [35].

Soil microorganisms are critical for the maintenance of functions in both natural and managed soils because they are involved in several key processes, such as decomposition of SOM, soil structure formation, the cycling of carbon, nitrogen, phosphorus, and sulfur, and toxin removal. Moreover, microorganisms are fundamental in promoting plant growth and in suppressing soil-borne plant diseases [36]. There is mounting evidence that healthy soils may promote the suppression of plant diseases, pests, and pathogens mediated by soil biodiversity through predation, competition, and parasitism [37]. There is confirmation that belowground plant mutualists can ameliorate the impacts of pollution on plant growth [38], and earthworms have been suggested as useful facilitators of ecosystem services in abandoned mining areas [39] all those again might contribute to regenerate agriculture and improve productivity.

In general, everything that we eat, drink, breathe, clothes that we wear, and materials that we use pass through soil and soil biodiversity over and over again. Healthy soil with soil biodiversity at the center of sustainability programs is capable of providing most ecosystem services and therefore achieving compliance with SDGs and human well-being through regenerative agriculture [12]. Half of all sustainable development goals zero hunger, good health and wellbeing, clean water and sanitation, affordable clean energy, responsible consumption and production, climate action, and life on land (SDG-2, 3, 6, 7, 12, 13 and 15, respectively) depend on soil and regenerative agriculture [12].

There are currently several lines being explored for agriculture making better use of enhanced soil biodiversity: going back to wild crop relatives and how do they make use of microbiomes and can those traits be restored in current crops? And studying wild plant species along successional gradients to unravel how plants may be productive in high-diversity soils. Considering soil biodiversity also requires considering traits, interactions, and network structure (so, not only numbers). Soil biodiversity as a nature-based solution to enhance sustainability is possible, but it takes two to tango as it requires crops that can handle these soils [34].

Studies show that Arbuscular Mycorrhizal Fungi (AMF) can alleviate both biotic and abiotic stresses since they can contribute to restoring degraded lands and ecosystems via artificial inoculation, while they improve access of nutrients to plants. They also can regulate abiotic and biotic stresses to plants such as drought, salinity stress, heavy metal phyto-accumulation, and protection against pathogens [40, 41]. AMF are promising soil microorganisms that improve soil health through their influences on plant photosynthesis [42], nutrient transfer [43], root exudation [44], osmotic potentials [45], soil bacteria interactions [46], and soil structural improvement and as a trade-off nutrient uptake, disease control and phytoremediation [47, 48].

Microorganisms provide us many ecosystems service that results in soil health and consequently can be related to soil productivity and regenerating agriculture. For instance, nitrogen-fixing bacteria associate with legume roots fixes large amounts of nitrogen that are of pivotal importance for plant productivity. And the soil biodiversity is an important indicator of soil health in agriculture management [49].

Above all, agriculture needs a healthy full human resource to feed the increasing human population globally. Therefore, consideration of the roles of soil biodiversity on the medical sector or human health is necessary to use the full potentials of soil biodiversity for regenerating the ecosystem and agriculture. Soil microorganisms have an immense potential for the pharmaceutical industry because historically the discovery of numerous new drugs and vaccines; from well-known antibiotics like penicillin to bleomycin using for treating cancer and amphotericin for fungal infections and therapeutic measures for treating and controlling diseases comes from soil organisms [50]. As the systems of soil like that of the human system, the management aspect of the soil should be in line with biology rather than focussing on industrial chemistry.

3.3 Soil biodiversity dynamics

Soils support highly abundant and diverse communities of organisms that show a broad array of life histories and functional traits, and they range in body size from a few micrometers for some bacteria to several meters in length in the case of some earthworms. The soil microbial community is largely dominated by bacteria and fungi that account for most of the belowground biomass, roughly equal to 0.6 to 1.1% of soil organic C [51], and represent a biodiversity pool with estimated species richness of tens of thousands per gram soil [52]. Despite the importance of soil microorganisms, little is known about their distribution in the soil or how microbial community structure responds to changes in land management (**Tables 1** and 2).

Taxon	Diversity per amount soil or area (taxonomic units indicated below)	Abundance (approximate)
Bacteria and Archaea	$100–9{,}000 \cdot cm^{-3}$	$4–20 \cdot 10^9 \cdot cm^{-3}$
Fungi operational taxonomic units	$200–235 \; m.g^{-1}$	$100 \; m.g^{-1}$
AMF (species)	$10–20 \; m^{-2}$	$81–111 \; m.cm^{-3}$
Protists sequence	$150–1{,}200 \; (0.25 \; g)^{-1}$	$10^4–10^7 \cdot m^{-2}$
Nematodes (genera)	$10–100 \; m^{-2}$	$2–90 \cdot 10^5 \; m^{-2}$
Enchytraeids	$1–15 \; ha^{-1}$	$12{,}000–311{,}000 \; m^{-2}$
Collembola	$20 \cdot m^{-2}$	$1–5 \; 10^4 \; m^{-2}$
Mites (Oribatida)	$100–150 \; m^{-2}$	$1–10 \; 10^4 \cdot m^{-2}$
Isopoda	$10–100 \; m^{-2}$	$10 \cdot m^{-2}$
Diplopoda	$10–2{,}500 \; m^{-2}$	$110 \cdot m^2$
Earthworms (Oligochaeta)	$10–15 \; ha^{-1}$	$300 \cdot m^{-2}$

Table 1.
Estimated diversity and abundance of soil taxa according to published work of Bardgett and van der Putten [53].

Organism size	Group	Known species	Estimated species	% described
	Vascular plants	350 700	400 000	88%
	Macrofauna			
	Earthworms	7 000*	30 000*	23%
	Ants	14 000	25 000–30 000	60–50%
	Termites	2 700	3 100	87%
	Mesofauna			
	Mites	40 000*	100 000	55%
	Collembalans	8 500*	50 000	17%
	Microfauna ad microorganisms			
	Nematodes	20 000–25 000*	1 000 000–10 000 000*	0.2–2.5%
	Protists	21 000*	7 000 000–70 000 000*	0.03–0.3%
	Fungi	97 000	1 500 000–5 100 000	1.9–6.5%
	Bacteria	15 000	>1 000 000	<1.5%

Asterisks indicate numbers of species that live in the soil.
Source: Orgiazzi et al. [54].

Table 2.
Known and estimated number of species of soil organisms and vascular plants organized according to size.

3.4 Biofertilizer and concern of soil biodiversity in Ethiopia

While field research on bio-fertilizers in Ethiopia dating back to the early 1980s by the Institute of Agricultural Research; bio-fertilizers did not become available for farmers until 2010. Later, the National Soil Testing Center (NSTC), Menagesha Biotech Industry (MBI) PLC, and the Ethiopian Institute of Agricultural Research (EIAR) have developed capacities to produce Rhizobia-based biofertilizers (**Table 3**). Currently, postgraduate students and different researchers in Ethiopia have played significant roles in research activities of Rhizobial inoculants collection, characterization, selection, evaluation and revealed the potential of the local isolates to serve as biofertilizers at a commercial level to increase the yield of different leguminous crops [56].

Jida and Assefa have collected 30 isolates of efficient nitrogen-fixing lentil-nodulating rhizobia from farmers' field soils in central and northern parts of Ethiopia and selected for symbiotically efficient ones, which possess plant growth-promoting characteristics. Under glasshouse conditions, they found characteristics such as IAA production in 36.7% and inorganic phosphate solubilization capacity in 16.7% [57].

Fekadu and Tesfaye [58] reported that *P. fluorescens* isolates showed plant growth-promoting (PGP) traits like phosphate-solubilization, siderophore (molecules that bind ferric iron with an extremely high affinity) production, hydrogen cyanide production, ammonia production, and indole acetic acid (IAA) production. Hence, these isolates have been used as biocontrol agents and plant growth-promoting rhizobacteria (PGPR). Similarly, Diriba, [59] reported that among wild Arabica coffee rhizosphere isolates, *Bacillus* and *Pseudomonas* spp. in particular showed remarkable inhibition against *Fusarium xylarioides, F. stilboides,* and *F. oxysporum* under in vitro conditions. The same author has also reported that a considerable number of wild Arabica coffee-associated rhizobacteria *(Pseudomonas, Bacillus, Azospirillum, and Rhizobium* produce siderophores.

The research outputs of Muluneh and Zinabu [60] revealed that the application of dried cyanobacteria on lettuce crop increased the number of leaves, leaf area, leaf length, fresh weight of the leaf, leaf dry weight, and the root dry weight of the lettuce by 159.5, 112.4, 80.8, 48, 137.5 and 110%, respectively, over their control. Tesfaye *et al.* [61] concluded that Azolla should be used as a biofertilizer for rice production in Ethiopia since it produces high biomass, is easy to manage and establish, increases the availability of macro and micronutrients (it scavenges K and recycles P and S), improves soil physical and chemical properties and fertilizer use efficiency, increases crop yield by 15–19% (by one incorporation) in Ethiopia and releases plant growth hormones and vitamins and does not attract rice pests. Some

Crop	Types of inoculant (rhizobia)	Crop	Types of inoculant (rhizobia)
Faba bean and Field pea	Rhizobia leguminosarum vicea	Common bean	*R. leguminosarum* phasoeli
Chickpea	Mesorhizobium cicer	Cowpea	*B. elkanii*
Soybean	*B. japonicum*	Groundnut	Rhizobium spp
Lentil	*R. leguminosarum*	Alfalfa	*E. meliloti*

Source: EIAR [55].

Table 3.
Rhizobia species in commercially available inoculants (biofertilizers) for legume crops in Ethiopia as of March 2014.

experimental works were done on the use of mycorrhizae on coffee production by Tadesse and Fassil [62] and, a promising result was obtained on the sufficiency of phosphorus particularly.

Although different studies have been undertaking on microbial inoculation trials of several pulse crops in Ethiopia, the knowledge and data regarding Ethiopian soil biodiversity are very limited. Of course, the country has a responsible institute (Ethiopian Biodiversity Institute) to ensure the country's biodiversity and the associated community knowledge for proper conservation and sustainable utilization [63]. However, the most focus is given to above-ground diversity, but the attention given for belowground diversity is less which leads to the presence of limited knowledge and data in soil biodiversity throughout the country.

In 2015 the "Ethiopia's National Biodiversity Strategy and Action Plan 2015-2020" were developed through the involvement of different stakeholders and higher officials. However, the attention and discussions given for soil biodiversity in the document as well as in the key not messages of higher officials look limited. Different state ministers who participated in the event was forwarded their message regarding Ethiopia's geographical, climatic, cultural, linguistic diversity and then about the above-ground diversity (plant, birds, mammals, fish ...) but not on the diversity under their feet [64]. This reflects that society in general and policymakers, in particular, have neglected soil biodiversity; no attention was given to the large biodiversity pool stored belowground.

Generally, although some research findings were done and doing on the use of soil microbes as a biofertilizer in Ethiopia mainly by the academic group, there are no confidential estimates on the number of species, taxonomic groups, ecological functions and services, and interactions among soil organisms so far in Ethiopia. Moreover, there is no exact data on the level of threats to soil microbial genetic resources of the country. However, all factors affecting the ecosystem, plant, and animal biodiversity are believed to affect directly or indirectly the soil biodiversity base of the country [65]. Therefore, collecting, identifying, conserving, and knowing the status of soil biodiversity genetic resources of the country will be a forthcoming major task.

3.5 Challenges of soil biodiversity

The landmark FAO state about soil biodiversity for food and agriculture [66], the first-ever launch last year highlighted that associate biodiversity species living in around production seasons, particularly microorganisms and invertebrates, has never been documented. In many cases, there is a limited understanding of ecosystem function and service and consequently, the contribution of specific soil biodiversity components to the production systems is poorly understood [67]. This discloses that due to the presence of knowledge gap and complex interaction of soil life, soil biodiversity is increasingly under threat which results in changes in the composition of soil communities and loss of species, as well as the benefits they provide to all life. Therefore, governments and society all need to better understand the complexity of the interaction regarding all elements of future agriculture and the soils to think about resilience and food systems.

As a whole, soil degradation by erosion, land-use change, climate change, soil pollution, salinization, and sealing all threaten soil biodiversity by compromising or destroying the habitat of the soil biota. Management practices that reduce the deposition or persistence of organic matter in soils, or bypass biologically mediated nutrient cycling, also tend to reduce the size and complexity of soil communities. For instance, land-use intensification results in fewer functional groups of soil biota with fewer and taxonomically more closely related species [68]. Intensive agriculture and sealing of fertile lands due to urbanization can cause declines in abundance and species of soil biodiversity, making soil food webs less diverse [10, 69]. Wagg *et al.* [70] were also investigated this given recent observation that soil biodiversity is declining and that soil communities are changing upon land-use intensification. They showed that soil biodiversity loss and simplification of soil community composition impair multiple ecosystem functions, including plant diversity, decomposition, nutrient retention, and nutrient cycling. Louwagie [71] and Mujtar *et al.* [72] asserted that soil biodiversity tends to be greater in undisturbed natural lands as compared to cultivated fields.

Deforestation can alter the structure of soil communities and decrease species richness (including natural predators and pollinators) and leading to homogenization. Consequently, the area will have a reduction of ecosystem resilience due to organism imbalance, which can favor pests and disease outbreaks [73–75]. In the findings of Migliorini *et al.* [76] and Hong *et al.* [77] heavy metal, pollution can shift communities to become dominated by a few taxa that can tolerate, or even thrive with, high levels of chemical inputs with corresponding decreases in taxa abundant in unpolluted soils.

The introduction of all kinds of invasive alien species has harmed the above-ground biodiversity and the native soil biodiversity. The effects of invasive species in soil biodiversity vary depending on the species trophic position. Many invasive soil species are related to agricultural pests while certain species are introduced as biocontrol agents. Another example is the introduction of non-native earthworms (which are ecosystem engineers), but their invasions can cause cascading effects that impact plant communities, forest, carbon sequestration, and wildlife [78, 79].

3.6 Soil biodiversity management

Soil biodiversity is part of the biological resources of the agroecosystems and must be considered in national and international management decisions. As indicated in the publication of Lijbert *et al.* [80] the main management options for soil biodiversity comprise no-tillage, crop rotation, and organic matter management. Protecting existing natural areas, restoring degraded habitats, employing regenerative agricultural practices, and implementation urban biodiversity are important practices that support and sustain diverse soil communities and the functions and services they provide. Hence, adopting the use of intercrops, rotations, appropriate tillage methods, maintenance of soil cover and the incorporation of crop residues into the soil management practices favor beneficial soil biodiversity [81, 82].

Global Soil Partnership (GSP) was established in December 2012 to enhance collaboration and synergy of efforts for sustainable soil management, and to protect biodiversity through sustainable soil management. The GSP supports soil biodiversity enhancement through monitoring soil biodiversity; maintaining or enhancing soil organic matter levels; the regulation of authorization and use of pesticides in agricultural systems; the use of nitrogen-fixing leguminous species; restoring plant biodiversity and crop rotation. All those activities lead to sustainable soil management and higher and more stable productivity [83]. Over the past few years, there has been an increased interest in organic farming practices, which could have benefits for soil biodiversity, particularly owing to the reduced use of pesticides [84].

There is a need to celebrate discoveries about life under our feet, as well as to integrate knowledge about soil biodiversity into international policies. Understanding how limitations to agricultural production at various levels (social, cultural, economic, political, agronomic, biological, environmental, edaphic, genetic) can be overcome is essential, to predict possible management options for the conservation of soil biodiversity and regenerative agriculture. To restore soil biodiversity, there is a need to think about ecosystem management, first, and store by diversity and its multi-functionality that what they are doing and how they are interacting with other species.

4. Conclusion

Soil biodiversity is highly linked with the ecosystem functions and services in the atmosphere, hydrosphere, lithosphere, and biosphere; all those do have their contribution and influence on land resources and agriculture. The soil organisms are contributing to climate mitigation and adaptation, water infiltration, and purification, soil health improvements and productivity, and playing countless roles through modification of conditions for the proper plant, animal, and human health all those are intern involved in regenerative agriculture. So, regenerative agricultural practices and soil biodiversity are evolved together and they are important components for future agricultural directions both in developed and developing countries.

However, there are knowledge and skill gaps in the area of soil biodiversity particularly on invertebrates and soil microorganisms to taxonomically classify and determine the complex interaction among themselves and the environmental factors. Therefore, more attention should be given to the discoveries of soil biodiversity and moving beyond academic circles particularly in the developing countries, to use them as a natured based solution for regenerative agriculture in the 21st century and to meet the 2030 sustainable development goals.

Acknowledgements

The author of this review thanks and highly appreciates the scientific contributions of the different authors, organizations, institutions, and webinar organizers cited in the text.

Author details

Mulugeta Aytenew
Department of Plant Science, College of Agriculture and Natural Resources, Debre Markos University, Ethiopia

*Address all correspondence to: mulugetaaytenew@gmail.com

IntechOpen

References

[1] Rhodes CJ. The imperative for regenerative agriculture. *Sci Prog.* 2017; 100(1):80-129. DOI: 10.3184/ 003685017X14876775256165

[2] Smith P, Cotrufo MF, Rumpel C, Paustian K, Kuikman PJ, Elliott JA, et al. Biogeochemical cycles and biodiversity as key drivers of ecosystem services provided by soils. *SOIL.* 2015;1:665-685. DOI: 10.5194/soil-1-665-2015

[3] Rhodes CJ. Feeding and healing the world: through regenerative agriculture and permaculture. *Science Progress.* 2012; 95(2):101-201

[4] HLPE (High Level Panel of Experts on Food Security and Nutrition). Sustainable agricultural development for food security and nutrition: what roles for livestock? Rome: A report by the High Level Panel of Experts on Food Security and Nutrition of the Committee on World Food Security; 2016

[5] Wassie, S.B., 2020. Natural resource degradation tendencies in Ethiopia: a review. *Environ Syst Res* 9 (33): (https:// doi.org/10.1186/s40068-020-00194-1

[6] Chapman G. What is regenerative agriculture? Southern Blue Regenerative. Armidale, NSW. 2019;2350

[7] https://www.regeneration-academy. org/mission.

[8] Christopher Johns, 2020. Growing Our Future – How Regenerative Agriculture Can Achieve Economies of Scale. Strategic Analysis paper, Future Directions International Pty Ltd. Hampden Road, Australia.

[9] Elevitch, Craig R.; Mazaroli, D. N.; Ragone, Diane, 2018. "Agroforestry Standards for Regenerative Agriculture". *Sustainability*10 (9): 3337. https://doi.org/10.3390/su10093337

[10] Anne Turbé, Arianna De Toni, Patricia Benito, Patrick Lavelle, Perrine Lavelle, Nuria Ruiz, Wim H. Van der Putten, Eric Labouze, and Shailendra Mudgal, 2010. Soil biodiversity: functions, threats and tools for policymakers. Bio Intelligence Service, IRD, and NIOO, Report for European Commission (DG Environment).

[11] FAO (Food and Agriculture Organization), ITPS (Intergovernmental Technical Panel on Soils), GSBI (Global Soil Biodiversity Initiative), CBD (Convention on Biological Diversity), and European Commission. 2020. State of knowledge of soil biodiversity – Status, challenges and potentialities, Summary for policy makers. Rome, FAO. https://doi.org/ 10.4060/cb1929en

[12] Bach EM, Ramirez KS, Fraser TD, Wall DH. Soil Biodiversity Integrates Solutions for a Sustainable Future. *Sustainability.* 2020;12(2662). DOI: 10.3390/su12072662w

[13] Chris Arsenault, 2014. Only 60 Years of Farming Left If Soil Degradation Continues. https://www. scientificamerican.com/article/

[14] Melvani K. Hand book for regenerative Agriculture. USAID, from the American people. 2016

[15] Rodale Institute. Regenerative Organic Agriculture and Climate Change. A Down-to-Earth Solution to Global Warming. Rodale Institute, 611 Siegfriedale Road, Kutztown. 2018

[16] André Leu, 2018, Reversing Climate Change through Regenerative Agriculture. https:// regenerationinternational.org/. Retrieved on January 25, 2021.

[17] Bradford MA, Carey CJ, Atwood L. Soil carbon science for policy and

practice. *Nat Sustain*. 2019;2:1070-1072
https://doi.org/10.1038/s41893-019-0431

[18] Poulton PR, Johnston AE, Macdonald AJ, White RP, Powlson DS. Major limitations to achieving 4 per 1000 increases in soil organic carbon stock in temperate regions: evidence from long-term experiments at Rothamsted Research, UK. *Global Change Biology*. 2018;24(6):2563-2584

[19] Dede Sulaeman and Thomas Westhoff, 2020. The Causes and Effects of Soil Erosion, and How to Prevent It. https://www.wri.org/blog/2020/01/

[20] López-Vicente M, Calvo-Seas E, Álvarez S, Cerdà A. Effectiveness of Cover Crops to Reduce Loss of Soil Organic Matter in a Rainfed Vineyard. *Land*. 2020;2020(9):230

[21] Cook RJ, Veseth RJ. Wheat Health Management. St. Paul, Minnesota: American Phytopathological Society Press; 1991

[22] Han-ming HE, Li-na LIU, Munir S, Bashir NH, Yi WANG, Jing YANG, et al. Crop diversity and pest management in sustainable agriculture. *Journal of Integrative Agriculture*. 2019;18(9): 1-1952

[23] Schreefel L, Schulte RPO, de Boer IJM, Pas Schrijver A, van Zanten HHE. Regenerative agriculture – the soil is the base. *Global Food Security*. 2020;26(2020):100404

[24] LaCanne and Lundgren Regenerative agriculture: merging farming and natural resource conservation profitably. *PeerJ*. 2018;6: e4428. DOI: 10.7717/peerj.4428

[25] Burgess PJ, Harris J, Graves AR, Deeks LK., 2019. Regenerative Agriculture: Identifying the Impact; Enabling the Potential. Report for SYSTEMIQ. 17. Bedfordshire. UK: Cranfield University; May 2019

[26] IPCC4 (Intergovernmental Panel on Climate Change)I, 2019: Climate Change and Land: an IPCC special report on climate change, desertification, land degradation, sustainable land management, food security, and greenhouse gas fluxes in terrestrial.

[27] Lunn-Rockliffe S, Davies MI, Willman A, Moore HL, McGlade JM, Bent D. Farmer Led Regenerative Agriculture for Africa. London: Institute for Global Prosperity; 2020

[28] FAO (Food and Agriculture Organization). The future of food and agriculture – Trends and challenges2017 Rome

[29] http://www.fao.org/3/u7260e/u7260e06.htm. Sustainable production systems. Retrieved on January 25, 2021.

[30] EPAT. (Extracorporeal Pulse Activation Technology), 1994. Pesticides and the agrichemical industry in sub-Saharan Africa. Environmental and Natural Resources Policy and Training Project. Report prepared for the Division of Food, Agriculture, and Resource Analysis, Office of Analysis, Research, and Technical Support, Bureau for Africa, U.S. Agency for International Development. Arlington, VA. USA: Winrock International Environmental Alliance.

[31] Kassie, M., Zikhali, P., Pender, J., & Köhlin, G., 2011. Environment for Development Initiative. http://www.jstor.org/stable/resrep14947: Retrieved January 25, 2021

[32] http://www.fao.org/3/Y4818E/y4818e07.htm. Policy options and strategies in natural resources management to enhance food security and broaden the livelihood base. Accessed on January 25, 2021.

[33] FAO (Food and Agriculture Organization) and ITPS

(Intergovernmental Technical Panel on Soils). Status of the World's Soil Resources (SWSR) – Main Report. Food and Agriculture Organization of the United Nations and Intergovernmental Technical Panel on Soils. Italy: Rome; 2015

[34] FAO (Food and Agriculture Organization) and GSP (Global Soil Partnership), 2020. Soil biodiversity: a nature-based solution Webinar. Soil Biodiversity/20/Report.

[35] Lemanceau P, Maron P-A, Mazurier S, Mougel C, Pivato B, et al. Understanding and managing soil biodiversity: a major challenge in agroecology. Agronomy for Sustainable Development, Springer Verlag/EDP Sciences/INRA. 2015;35(1):67-81

[36] Garbeva P, van Veen JA, van Elsas JD. Microbial diversity in soil: selection of microbial populations by plant and soil type and implications for disease suppressiveness. *Ann Rev Phytopathol*. 2004;42(1):243-270

[37] Susilo FX, Neutel AM, van Noordwijk M, Hairiah K, Brown G, Swift MJ. Soil biodiversity and food webs. In Below-Ground Interactions in Tropical Agro ecosystems: Concepts and Models with Multiple Plant. 2004

[38] Glassman SI, Casper BB. Biotic contexts alter metal sequestration and AMF effects on plant growth in soils polluted with heavy metals. *Ecology*. 2012;93:1550-1559

[39] Boyer S, Wratten SD. The potential of earthworms to restore ecosystem services after opencast mining—a review. *Basic Appl. Ecol*. 2010;11: 196-203

[40] Arafat Abdel Hamed Abdel Latef, Abeer Hashem, Saiema Rasool, Elsayed Fathi Abd_Allah, Alqarawi A.A., Dilfuza Egamberdieva, Sumira Jan, Naser A. Anjum, and Parvaiz Ahmad, 2016.

Arbuscular Mycorrhizal Symbiosis and Abiotic Stress in Plants: A Review. J. Plant Biol. (2016) 59:407-426 DOI 10.1007/s12374-016-0237-7

[41] Diagne N, Ngom M, Djighaly PI, Fall D, Hocher V, Svistoonof S. Roles of Arbuscular Mycorrhizal Fungi on Plant Growth and Performance: Importance in Biotic and Abiotic Stressed Regulation. *Diversity*. 2020;12:370. DOI: 10.3390/d12100370

[42] Zhu XQ, Wang CY, Chen H, Tang M. Effects of arbuscular mycorrhizal fungi on photosynthesis, carbon content, and calorific value of black locust seedlings. *Photosynthetica*. 2014;52:247-252

[43] Azcón-Aguilar C, Barea JM. Arbuscular mycorrhizas and biological control of soil-borne plant pathogens— an overview of the mechanisms involved. *Mycorrhiza*. 1997;6(6): 457-464

[44] Gupta MM, Aggarwal A Asha. 2018. From Mycorrhizosphere to Rhizosphere Microbiome: The Paradigm Shift. In Root Biology (pp. 487-500). Springer, Cham

[45] Chen LQ. Sweet sugar transporters for phloem transport and pathogen nutrition. *New Phytol*. 2014;201:1150-1155

[46] Toro M, Azcon R, Barea J. Improvement of Arbuscular Mycorrhiza Development by Inoculation of Soil with Phosphate-Solubilizing Rhizobacteria to Improve Rock Phosphate Bioavailability ((sup32) P) and Nutrient Cycling. *Appl. Environ. Microbiol*. 1997;63(11): 4408-4412

[47] Jacott CN, Murray JD, Ridout CJ. Trade-offs in arbuscular mycorrhizal symbiosis: disease resistance, growth responses and perspectives for crop breeding. *Agronomy*. 2017;7(4):75

[48] Yang Y, He C, Huang L, Ban Y, Tang M. The effects of arbuscular

mycorrhizal fungi on glomalin-related soil protein distribution, aggregate stability and their relationships with soil properties at different soil depths in lead-zinc contaminated area. *PLoS ONE.* 2017;12(8):e0182264

[49] SHI (Soil Health Institute), 2017. North American Project to Evaluate Soil Health Measurements. https://soilhealthinstitute.org/

[50] Luisa AD, Kuenzi AJ, Mills JN. Species diversity concurrently dilutes and amplifies transmission in a zoonotic host-pathogen system through competing mechanisms. *PNAS.* 2018;115 (31):7979-7984

[51] Fierer N, Strickland MS, Liptzin D, Bradford MA, Cleveland CC. Global patterns in belowground communities. *Ecol. Lett.* 2009;12:1238-1249

[52] Roesch LF, Fulthorpe RR, Riva A, Casella G, Hadwin AKM. Pyrosequencing enumerates and contrasts soil microbial diversity. *ISME J.* 2007;1:283-290

[53] Bardgett R, van der Putten W. Belowground biodiversity and ecosystem functioning. *Nature.* 2014; 515:505-511

[54] Orgiazzi A, Bardgett RD, Barrios E, BehanPelletier V, Briones MJI, Chotte J-L, et al. Global soil biodiversity atlas. Luxembourg, European Commission: Publications Office of the European Union; 2016

[55] EIAR (Ethiopian Institute of Agricultural Research). Rhizobia-based bio-fertilizer; Guidelines for smallholder farmers. Ethiopia: Addis Ababa; 2014

[56] Keneni G. Overview of rhizobial inoculants research and biofertilizer production for increased yield of food Legumes in Ethiopia. EJCS. 6 (3) Special Issue (1). 2018

[57] Jida M, Assefa M. Phenotypic and plant growth-promoting characteristics of Rhizobium leguminosarum viciae from lentil growing areas of Ethiopia. *Afr J Microbiol Res.* 2011;5(24):4133-4142

[58] Alemu F, Alemu T. Pseudomonas fluorescens Isolates used as a Plant Growth Promoter of Faba Bean (Vicia faba) in Vitro as Well as in Vivo Study in Ethiopia. *American Journal of Life Sciences,* 3(2):100-108. doi: 10.11648/j.ajls.20150302.17. 2015

[59] Muleta D. Microbial Inputs in Coffee (Coffea arabica L.) Production Systems, Southwestern Ethiopia: Implications for Promotion of Biofertilizers and Biocontrol Agents. Doctoral thesis. Acta Universitatis Agriculturae Sueciae: 117. 2007

[60] Menamo M, Wolde Z. Effect of Cyanobacteria Application as Biofertilizer on Growth, Yield and Yield Components of Romaine Lettuce (Lactuca sativaL.) on Soils of Ethiopia. *ASRJETS.* 2013;4(1):50-58

[61] Feyisa T, Amare T, Adgo E, Selassie YG. Symbiotic Blue-Green Algae (Azolla): A Potential Biofertilizer for Paddy Rice Production in Fogera Plain, Northwestern Ethiopia. *Eth. J. Sci & Technol.* 2013;6(1):1-11

[62] Sewnet TC, Tuju FA. Arbuscular mycorrhizal fungi associated with shade trees and Coffea arabica L. in a coffee-based agroforestry system in Bonga, Southwestern Ethiopia. *Afrika Focus.* 2013;26(2):111-131

[63] https://www.ebi.gov.et.

[64] Ethiopian Biodiversity Institute. Ethiopia's National Biodiversity Strategy and Action Plan 2015–2020. Government of the Federal Democratic Republic of Ethiopia. Ethiopia: Addis Ababa; 2015

[65] Anton M. Breure, 2004. Soil biodiversity: measurements, indicators,

threats and soil functions. International Conference on soil and compost eco-biology: September 15th – 17th 2004, León - Spain

[66] FAO (Food and Agriculture Organization), 2019. The State of the World's Biodiversity for Food and Agriculture, J. Bélanger & D. Pilling (eds.). FAO Commission on Genetic Resources for Food and Agriculture Assessments. Rome. 572 pp.

[67] FAO (Food and Agriculture Organization), 2021. Concept Note: Global Symposium on Soil Biodiversity (GSOBI21) "Keep soil alive, protect soil biodiversity" held on 2 – 5 February 2021 – Virtual meeting (Zoom platform), FAO headquarters, Rome, Italy

[68] Tsiafouli MA, Thébault E, Sgardelis S, De Ruiter PC, Van der Putten WH, Birkhofer K, et al. Intensive agriculture reduces soil biodiversity across Europe. *Global Change Biology*. 2014;21:973-985

[69] Tibbett M, Fraser TD, Duddigan S. Identifying potential threats to soil biodiversity. *PeerJ*. 2020;8:e9271. DOI: 10.7717/peerj.9271

[70] Wagg C, Franz Bender S, Widmer F. Soil biodiversity and soil community composition determine ecosystem multifunctionality. *PNAS*. 2014;111(14):5266-5270

[71] Louwagie G. Final report on the project 'Sustainable Agriculture and Soil Conservation (SoCo)'. Luxembourg: European Communities; 2009

[72] El Mujtar V, Muñoz N, Prack Mc Cormick B, Pulleman M, Tittonell P. Role and management of soil biodiversity for food security and nutrition; where do we stand? *Glob. Food Sec.* 2019;20:132-144

[73] Crowther TW, Maynard DS, Leff JW, Oldfield EE, McCulley RL,

Fierer N, et al. Predicting the responsiveness of soil biodiversity to deforestation: a cross-biome study. *Global change biology*. 2014;20(9): 2983-2994

[74] Franco ALC, Sobral BW, Silva ALC, Wall DH. Amazonian Deforestation and Soil Biodiversity. *Conservation Biology*. 2018;33:590-600

[75] Kroeger ME, Delmont T, Eren AM, Meyer KM, Guo J, Khan K, et al. New biological insights into how deforestation in amazonia affects soil microbial communities using metagenomics and metagenomeassembled genomes.*Frontiers in microbiology*,9,1635. 2018

[76] Migliorini M, Pigino G, Caruso T, Fanciulli PP, Leonzio C, Bernini F. Soil communities (Acari Oribatida; Hexapoda Collembola) in a clay pigeon shooting range. *Pedobiologia*. 2005;49: 1-13

[77] Hong C, Si Y, Xing Y, Li Y. Illumina MiSeq sequencing investigation on the contrasting soil bacterial community structures in di_erent iron mining areas. *Environ. Sci. Pollut. Res.* 2015;22: 10788-10799

[78] Ehrenfeld JG, Scott N. Invasive Species and the Soil: Effects on Organisms and Ecosystem Processes. *Ecological Applications*. 2001;11(5): 1259-1260

[79] Rai PK, Singh JS. Invasive alien plant species: Their impact on environment, ecosystem services, and human health. *Ecol Indic*. 2020;111: 106020

[80] Brussaard L, de Ruiter PC, Brown GG. Soil biodiversity for agricultural sustainability. *Agriculture, Ecosystems and Environment*. 2007;121: 233-244

[81] Brooker RW, Bennett AE, Cong WF, Daniell TJ, George TS,

Hallett PD, et al. Improving intercropping: a synthesis of research in agronomy, plant physiology and ecology. *New Phytologist*. 2015;206(1): 107-117

[82] FAO (Food and Agriculture Organization), 2003. Biological management of soil ecosystems for sustainable agriculture. Report of the International Technical Workshop organized by EMBRAPA-Soybean and FAO, Londrina, Brazil, 24–27 June 2002. Rome. (Available at http://www.fao. org/docrep/006/y4810e/ y4810e00.htm).

[83] FAO (Food and Agriculture Organization) and ITPS (Intergovernmental Technical Panel on Soils). Protocol for the assessment of Sustainable Soil Management. FAO: Rome; 2020

[84] Tahat MM, Alananbeh KM, Othman YA, Leskovar DI. Soil Health and Sustainable Agriculture. *Sustainability*. 2020;12:4859

Chapter 11

Soil Biodiversity and Root Pathogens in Agroecosystems

María del Pilar Rodríguez Guzmán

Abstract

Soil ecosystem is a living and dynamic environment, habitat of thousands of microbial species, animal organisms and plant roots, integrated all of them in the food webs, and performing vital functions like organic matter decomposition and nutrient cycling; soil is also where plant roots productivity represent the main and first trophic level (producers), the beginning of the soil food web and of thousands of biological interactions. Agroecosystems are modified ecosystems by man in which plant, animal and microorganisms biodiversity has been altered, and sometimes decreased to a minimum number of species. Plant diseases, including root diseases caused by soil-borne plant pathogens are important threats to crop yield and they causes relevant economic losses. Soil-borne plant pathogens and the diseases they produce can cause huge losses and even social and environmental changes, for instance the Irish famine caused by *Phytophthora infestans* (1845–1853), or the harmful ecological alterations in the jarrah forests of Western Australia affected by *Phytophthora cinnamomi* in the last 100 years. How can a root pathogen species increase its populations densities at epidemic levels? In wild ecosystems usually we expect the soil biodiversity (microbiome, nematodes, mycorrhiza, protozoa, worms, etc.) through the trophic webs and different interactions between soil species, are going to regulate each other and the pathogens populations, avoiding disease outbreaks. In agroecosystems where plant diseases and epidemics are frequent and destructive, soil-borne plant pathogens has been managed applying different strategies: chemical, cultural, biological agents and others; however so far, there is not enough knowledge about how important is soil biodiversity, mainly microbiome diversity and soil food webs structure and function in the management of root pathogens, in root and plant health, in healthy food production, and maybe more relevant in the conservation of soil as a natural resource and derived from it, the ecosystem services important for life in our planet.

Keywords: soil biodiversity, soilborne plant pathogens, soil food webs, ecological interactions, plant pathogens management

1. Introduction

Soil ecosystem is a living and dynamic environment, habitat of thousands of microbial species, animal organisms and plant roots, integrated all of them in the food webs, and performing vital functions like organic matter decomposition, nutrient cycling and release, promote plant growth, receive, hold and release water, transfer energy in the detritus food chain, and act as an environmental buffer [1–3]; soil is also where plant roots productivity represent the main and first trophic level (producers) [4], the beginning of the soil food web and of thousands of biological interactions [5, 6]. In Agroecosystems, the different activities practiced by man will affect all the biological processes carried out above- and belowground, including soil biodiversity and soil food webs [7], depending of the kind of agroecosystem (traditional or intensive), the geographical region, the crop management, and social and economic interests. Human societies have developed several kinds of agroecosystems, from traditional/subsistence, multicropping, to intensive and highly technified crops [8–10] however, plant pathogens and pest diseases are a common component of the agroecosystem, and by some degree all of them are affected [11]. In this sense, a relevant and unavoidable problem is the soil degradation and contamination in agroecosystems; plant diseases caused by soilborne plant pathogens (SBPP) are of great importance because most of the strategies applied for their control are directed to the soil [12]. Management of SBPP and diseases require a broader view and a thorough ecological knowledge of the soil ecosystem, considering the improvement and conservation of the soil biodiversity and the soil food web structure and function, and studying the soil as a dynamic ecosystem in time and space. Plant pathologist and agronomist must know about the importance of the soil ecosystem, its biodiversity, the different and multiple functions soil organisms perform, how every and all soil organisms are connected through different relationships established as a result of natural selection forces, and how they have evolved throughout the time [13, 14]. And maybe most important is to understand how a disturbance or stress factor imposed on soil organisms may have a cascade effect in the whole processes and functions of the soil ecosystem. In this work we are talking about the biological soil diversity and functions, the soil food web, the ecological interactions among species and how they are important in the management of root pathogens and the diseases they cause. It seems there is an urgent need for redesigning and developing sustainable agroecosystems where soilborne plant diseases and pathogens be analyzed under an integral knowledge with the application of plant pathology, plant disease epidemiology, and ecological and evolutionary principles [15]. Even more if the agroecosystems of interest involve soilborne plant pathogens and diseases, studies must comprehend the important role that soil biodiversity [16] and its multiple interactions and relationships play in the regulation of the SBPP, and how this regulation is expressed through the soil food web (structure and function) and through other complementary relationships.

2. Soil biodiversity

The soil is an ecosystem, a living system where interplay mineral and organic materials. Soils are built up through millions of years, from the parent rock layer to the small sand, lime and clay particles derived from physical and chemical intemperization processes, and from biological activities which perhaps be the most important factor in the soils formation. Of the total soil components, organic material represents 5 to 10%, from this percentage 10 to 20% is the active fraction, from the active organic fraction only 20 to 40% are living organisms, and from these 50% are fungi, 30% are bacteria and actinomycetes, 10% are yeasts, algae, protozoa, and nematodes, and 10% are fauna. This means, the active microbial biomass represents 90% of the soil living organisms. Another essential component in soil biodiversity is the soil fauna which is divided in function of their size (mm) into micro-, meso- and macrofauna. Microfauna includes Protozoa, Rotifers from 0.005 to 0.2 mm; mesofauna is composed by nematodes, arthropods, enchytraeids, mites, springtails from 0.2 to 10 mm; macrofauna include animals like insects from 10 to 20 mm; and megafauna includes earthworm (≥ 20 mm), macroarthropods and small mammals (cm) [2, 3, 17, 18].

Plant roots are another fundamental component of the soil biodiversity, they represent the primary source of organic matter in soil, and the amount of root production is relevant in the process of decomposition and cycling into organic matter in soils. Roots are the subterranean organ of the plant, and they fix them to the soil. Plants take water and nutrients from soil through the roots and the root vascular system transport them to the upper parts of the plant; in some cases, roots are a storage organ [19]. Root exudates and rhizosphere are relevant components for root functioning, and they are essential for all the biological and soil microbial activities like attraction of mutualist symbionts and pathogenic microbes, release and cycling of nutrients, allelopathic processes [5, 6] and also for physical and chemical soil characteristics such as soil aggregation and structure, cation exchange capacity and pH [20]. Plant roots represent the first trophic level (autotrophs entities) in the soil trophic web, from which microbes and small fauna obtain their nutrients and energy [18].

Soil-borne plant pathogens (SBPP) are a component of the soil ecosystem, and also members of the soil biodiversity; these microorganisms live part of their life in soil and in the plant rhizosphere, but they also infect and damage the plant roots from which they feed; in some way plant roots are their habitat. SBPP include organisms from bacteria, fungi, oomycetes, nematode, protozoa groups and meso-biotic entities like virus and viroids [11, 21].

3. Importance and function of soil biodiversity ¿who are here and what they do?

3.1 Bacteria and fungi

Bacteria and fungi participate in the rock intemperization (degradation) through their biochemical enzymatic activity; these microorganisms initiate the soil formation, and with it the important process of mineral transformation and nutrient liberation [22]. As soil is formed and deeper, it is possible for other microorganisms and larger organisms like protozoa, rotifers, nematodes, worms, small arthropods, and also spores of bryophytes, moss, ferns, and mycorrhizal fungi to arrive and to establish. When soil has enough deep and biological activity is adequate for higher plant seeds to germinate and their root systems to interact with the soil microbiome, it begins the formation of a plant community and in time, with the integration of higher animals and through a successional process, the establishment of a biome after hundreds and millions of years.

3.2 Soil bacteria

Certain groups of the soil bacteria community participates in the N cycle which involve four stages or reactions: a) Nitrogen fixation carried out by nitrogen-fixing bacteria (e.g. *Rhizobium*, *Azotobacter*, *Bradyrhizobium*), b) Ammonification performed by ammonifying bacteria or decomposers bacteria (*Bacillus subtilis*, *Pseudomonas fluorescens*), c) Nitrification accomplished by nitrifying bacteria (e.g. *Nitrosomonas*: NO_2 nitrite, *Nitrobacter*: NO_3 nitrate), and d) Denitrification realized by denitrifying bacteria (some species of *Serratia* and *Pseudomonas*) [23, 24]. It must be noted that bacteria also participate in the C biogeochemical cycle, and they play a crucial role in the regulation of C and N cycles during biological soil crust succession in arid and semi-arid ecosystems [24]. It is important to indicate that some of these bacteria species besides to participate in the Nitrogen cycle, like *Bacillus subtilis*, *B. amyloliquefasciens*, *Bradyrhizobium* (Nod Factors), they also are involved in a complex network of signaling pathways mediated by plant hormones like jasmonic acid, ethylene and salicilic acid and in the release of volatile organic compounds which trigger the induced systemic resistance (ISR) and the acquired systemic resistance (ASR) in plants, in response to the presence of both beneficial microbes and plant pathogens invasion and infection, and also to insect attack [25–30]. Some bacteria specific strains also produce different antibiotics, e.g., *Pseudomonas fluorescens* F113 produces 2,4-diacetyl-phloroglucinol against *Pythium* spp. [31], *Bacillus amyloliquefasciens* FZB42 produces bacillomycin and fengycin against *Fusarium oxysporum* [32]; enzymes, and siderophores e.g., *Pseudomonas putida* WCS358 against *Pseudomonas syringae* pv. tomato and *Fusarium oxysporum* f. sp. *raphani* [33], which are important components in their performance as biological control agents against plant pathogens [7, 34]. Another important activity performed by some soil bacteria is enhancing the plant growth through the production of growth regulators (hormones) like auxins (Indol Acetic Acid, IAA), cytokinins, gibberellins, or ethylene, and these bacteria are known as Plant-Growth Promoting Bacteria (PGPB) [35]; examples of bacteria species involved in this activity are *Pseudomonas fluorescens*, *P. putida*, *P. gladioli*, *Bacillus subtilis*, *B. cereus*, *B. circulans*, and bacteria in the genre *Azospirilum*, *Serratia*, *Flavobacterium*, *Alcaligenes*, *Klebsiella* and *Enterobacter* [7].

3.3 Soil fungi

Fungi, organisms in the Kingdom Fungi, perform different and important functions in soils, they are organic matter decomposers due to their great ability to degrade complex substrates of plant origin [36]. Fungi participate in the mineral degradation and in the release and cycling of nutrients; they are also involved in the C and N cycling [24, 36, 37]. These microorganisms are vital for soil functioning because most carbon in our planet is stored in rocks and sediments [38]. They also contribute to the soil particles aggregation and soil structure because of their filamentous form and exudates [39]. In soils exist a complex and diverse fungal community widely distributed [40]. This fungal community is composed by different and important functional groups. One of the most studied is the mycorrhizal fungi, which exists as mutualist symbionts in most of the plant species in natural ecosystems and it has a long evolutionary history [41, 42]; some of them develop a net of hyphae external to the roots and growing intercellularly in the root cortex, they are ectomycorrhiza; but other mycorrhiza can penetrate the roots and establish intracellularly in the cortex cells forming small structures called arbuscules, they are named arbuscular mycorrhiza or endomycorrhiza. Both kind of mycorrhiza help the plant host in the uptake of nutrients from soil and protection against pathogens. Arbuscular mycorrhizae play a central role in the Phosphorus cycle, but they are equally important in the Nitrogen cycle [43, 44]. Mycorrhiza hyphae link the plants roots with the soil particles, interconnect directly the root systems of two different individual plants, and they also interact with different kind of soil microbes (synergistic and antagonistic); even more, there is evidence of the extra radical mycorrhizal hyphae associated with symbiont bacteria (hyper symbionts) for acquisition of C [45]; therefore, ecto and endomycorrhizae fungi play important functions in the physiology, ecology and evolution of their host plants.

Fungal Endophytes are other important functional group of fungi, and they enter and live inside the plants [25, 42]. Here, endophytes are defined as those microorganisms (bacteria, fungi, virus) which live their life cycle or part of it inside a plant, within asymptomatic tissues, performing and promoting a beneficial functioning in the plant host, and enhancing its fitness in plant communities by conferring abiotic and biotic stress tolerance; therefore, this relationship has ecological and evolutionary importance [42, 46, 47]. Endophyte organisms can be found in different plant organs like roots, stems, leaves, reproductive organs (e.g., vanilla flower ovaries) and fruits (e.g., vanilla pods) [42, 48]. Endophytic fungi participate in different plant functions, some of them enhance plant growth and nutrition and are referred as Plant-Growth Promoting Fungi (PGPF) [25], and they also strengthen plant defense against pathogens and insects below- and above-ground [46, 47]. Endophyte fungi control plant pathogens attack through different processes: niche exclusion, antibiosis, predation, mycoparasitism and ISR induction [25]; it is also possible to find hypovirulent pathogen isolates which will control more virulent isolates as happens with *Monosporascus cannonballus* against monosporascus root rot vine decline [49], or with *Fusarium oxysporum* strain Fo47 [50, 51]. Endophyte fungi can express simultaneously more than one control mechanism against plant pathogens as it was showed with *Trichoderma* isolates that besides attack directly *Botrytis cinerea*, also induced systemic resistance to this pathogen [52]. Other fungal endophytes additionally of increasing plant biomass, confer drought tolerance, and produce chemicals that are toxic to animals like insects [46, 47], birds and small mammals and decrease herbivory [53]. Certain endophytic fungi have an important role in physiological and biochemical aspects during development of flower and chemical compounds, as is the case of *Vanilla planifolia* compound vanillin [48]; researchers have found that fungal endophytes inoculum from soil get into the roots through rhizosphere, but other endophytes come from the fungal airborne inoculum and enter into the flower ovaries, and later in the vanilla pods, participating in the vanillin process and therefore in the vanilla flavor [48]. Endophyte fungi described as PGPF include important genera like *Fusarium*, *Trichoderma*, *Aspergillus*, *Penicillium*, *Colletotrichum*, *Cylindrocladium*, and others; some of them are nonpathogenic or hypovirulent strains of plant pathogenic fungi [25, 42].

Some soil fungi are pathogens of other microorganisms like bacteria [54], fungi and nematodes [55], plants [11], insects and arthropods [47]; for instance, *Metarhizium anisopliae* and *Beauveria bassiana* are endophytic and pathogenic in insects, while *Paecelomyces lilacinus* is endophytic and pathogenic in nematodes [55]; these fungi take part in the regulation of their hosts populations and they are of relevance in the biological control and management of agroecosystems [46]. As mentioned above soil fungi are included in several functional groups: decomposers, mutualists (mycorrhiza), endophytes, pathogens, parasites, and every one of these activities are of great relevance for the soil ecosystem function.

3.4 Soil virus

Viruses are considered entities between living and non-living state, quasiorganisms; they are composed by RNA or DNA molecules contained within protein capsids, and they are mainly known as pathogens in plants, animal, and the human being, causing important diseases. However, and fortunately, with the help of the molecular biology methods, in the last 10 years there has been an unprecedent interest and research about virus diversity and functions in different environments: marine and soils [56, 57]. Knowledge about soil viruses are just beginning, very little is known about their ecology in soils [2, 56, 58]. However, it is suggested they participate in the biogeochemical cycling of Carbon [59], as well as in short-term adaptation and long-term evolution of microbes [2], through their infection like bacteriophage on beneficial bacteria (Rhizobia) and soilborne plant pathogens (bacteria, fungi, virus, nematodes and other soil organisms) [56]; they also perform horizontal gene transfer (transduction) among bacteria

[60]. Viruses impact the evolution and ecology of their plant hosts, and they seem to have a mutualistic relationship rather than a pathogenic one under experimental laboratory conditions [57]. On the other hand, plant pathogen viruses cause great economic and yield losses in agroecosystems where they are vectored by insects (aphids, white flies, trips, etc.), but they also are transmitted by mechanical ways. There are just few soilborne plant pathogenic viruses known so far, and they are transmitted by fungi and nematodes [61, 62].

3.5 Soil protozoa

Protozoa are other important component in the soil ecosystem and in the food web. The free-living protozoa feed from microbes like bacteria and fungi (non-pathogenic and pathogenic) [63], and also from algae; they are included in four groups: flagellates, naked amoebae, testacea amoebae and ciliates. They contribute to the regulation of these microbes population densities and dynamic. Protozoa also play an important role in the nutrient turnover [64].

3.6 Soil nematodes

Soil nematodes are some of the most abundant invertebrate animals in soils, they often reach densities of 1 million/m^2; they are worm-like microorganisms and live in water films or water-filled pore spaces in soils [64]. Many kinds of nematodes are found in the rhizosphere of roots and root hairs because of the rich exudates. They help to accelerate organic matter decomposition when they graze on bacteria, fungi, and plant residues [18]. Nematodes biological characteristics like structure, physiology, diverse reproductive patterns, and adaptability help them to inhabit many and different environments [65].

3.7 Earthworms

Earthworms participate in the fragmentation, breakdown, and incorporation of the soil organic matter, affecting its physical and chemical characteristics, and in turn other soil biota organisms [2]. They also affect positively soil structure helping in pores formation and particles aggregation, contributing to the soil aeration and better water distribution. Earthworms play an important role in C and N cycling [66]. Ecologically, earthworms promote diversity of fungal species and oribatid mites through their casts from where they feed, and through reducing competition between fungal species [67].

3.8 Soil arthropods

Arthropods in soil, are other important component in soil biodiversity. After microbes and Protozoa, microarthropods play a very important role in soil activities, they participate in the organic soil matter decomposition, nutrient release and cycling, but they also enhance plant growth and the expression of induced systemic resistance to pests in plants [68]; participate in the secondary seed dispersal of higher plants and dispersal of sperm in lower plants like mosses [69]; they are involved in the regulations of population densities of bacteria and fungi including plant pathogenic organisms and decomposition of agrochemicals [17, 70]. Microarthropods like collembola, protura, diplura, isopoda and others are also components in the soil food web, and they have been included as indicators of soil health and soil disturbance because they live a sedentary life and express the habitat conditions better than those organisms with a high dispersal capacity [18, 71, 72].

3.9 Soil-borne plant pathogens (SBPP)

Different groups of soil microorganisms bacteria, fungi, nematodes, protozoa and entities like virus and viroids may also act as plant root pathogens, and they received the name of soil-borne plant pathogens. These plant pathogens damage all kind of plants in the different botanical taxa in both natural and worldwide managed ecosystems; however, in agroecosystems their damages have economical relevance due to the resulted crop yield losses [73, 74]. However, SBPP like other plant pathogens play important roles in the structure, function, and diversity of natural plant communities [75–78]; they also are important in the evolution of the plant host-pathogen relationship [13, 79].

We can see, microorganisms have developed multiple, diverse and vital relationships through time and space with all other organisms in the soil ecosystem including plant roots, insects, and animals, and maybe most of these relationships have been developed in response to nature selection forces throughout an evolutionary time. It is necessary to mention that soil type [80] and soil management [81] have an important influence on the diversity and structure of soil microbial diversity, and in other soil microorganisms such as protozoa, rotifers, nematodes, microarthropods, mites, and of course in worms, ants, termites and small mammals [18]. However, plant roots, the principal C source in soils, and their rhizosphere exudates determine in relevant way, the spatial structure and diversity of the soil microbial and microorganisms community [82].

4. Trophic webs and complementary or interference interactions in soil "not everything in life is food"

Since the beginning of soil formation from the parental rock, biological activity is fundamental, many biological interactions are established, and they initiate the soil trophic web. At the beginning, trophic webs may be simple and with few components, but as root biomass and their exudates increase in amount and different types, the soil trophic webs are more complex in their biological diversity, structure and functions. Plant roots are key components in soil function, they provide most of the organic matter to the soils [4], they are the first trophic level in the soil food web and represent the autotroph organisms (photosynthesizers) from which heterotrophs organisms in the next trophic levels obtain their food and energy [83]. Plant roots and specifically the rhizosphere region have a transcendental role in the dynamic of soil microbial activities through the development and release of rhizodeposits [6, 84]. Rhizodeposits include sloughed-off root cap and border cells, mucilage, and exudates. The exudates are made up of organic acids, amino acids, proteins, fatty acids, enzymes, sugars, phenolic metabolites and other metabolites which are used by microorganisms [84, 85]. Most of the root exudates are released at the root cap and the meristematic zone behind the root cap; therefore, these regions are considered important for determining the temporal and spatial activity and distribution of the microbial communities [82, 84, 85].

Bacteria and fungi together with protozoa, plant pathogenic nematodes and fungi, are in the second trophic level of the soil food web, they are heterotrophs and they may function as decomposers, mutualists, pathogens, parasites and root-feeders, and they are food for the third trophic level (heterotrophs) that includes also nematodes and protozoa (bacterivorous and fungivorous), and microarthropods which work as shredders, predators and grazers; organisms from this level are food for the fourth and fifth trophic levels (heterotrophs) which include higher level predators like arthropods, small mammals, birds [83].

Soil protozoa is a group of microorganisms which has not been studied so deep and frequently like bacteria or fungi, and their role in the soil food web is sometimes considered only like predators and grazers of microbes; however, they interact with the root systems and bacteria in several and particularly important modes. Bonkowski and Brandt [86] worked with an Amoebae, specifically with *Acanthamoebae* sp. which is considered the most common soil free-living protozoon, and they found this amoebae have a positive effect on root elongation and branching in interaction with rhizobacteria, and mention that "Protozoa function as bacteria-mediated mutualists promoting plant growth by hormonal feed-back mechanisms and nutrient effects based on nutrient release from grazed bacterial biomass"; these bacteria may also be involved in the different phases of the soil nitrogen cycle. All these activities occur in the soil microbial loop [87], as a relevant component of the soil food web.

Soil nematodes are considered important component in the soil food web; their soil communities are usually large and species-rich, with different functional groups (bacterivores, fungivores, herbivores, omnivores, predators, parasites and pathogens), located in different trophic levels: root feeders nematodes in second trophic level (decomposers), fungi and bacterial feeder nematodes in third trophic level (grazers), nematode and protozoa predators in fourth trophic level (higher level predators), in this way nematodes participate in the regulation of soil microbial communities and indirectly in the flux of plant nutrients. Because of their biological characteristics (they are ubiquitous, abundant and diverse), they respond soon to changes in the soil environment, and for this reason they have been considered as important indicators of the soil health [18, 70, 88, 89].

Plant pathogens and therefore soilborne plant pathogens are also an important component inside the soil food web and the ecosystems [90, 91]. They can be considered as microherbivores because they feed over root systems, and later they are food for the next trophic level, protozoa, nematode, rotifers, mites, and microarthropods.

4.1 Plant root diversity and their effects in soil diversity and soil food web

Plant roots are the principal biomass and C source in soils; as roots and their exudates grow, they die and are decomposed by soil microorganisms and incorporated into the soil organic matter [4]. Plant roots are of many different types, lengths and architecture with a main and secondary roots and root hairs [92]. Roots produce different kind of exudates, and this is influenced by the plant species, soil physical and chemical conditions, soil temperature and moisture [82, 93]; however, root exudates are also affected by the rhizosphere microbial community [82]. Root exudates supply nutrients, they prevent invasion by other plant species (allelopathy), they function also like especially important chemical signals for attracting symbionts (chemotaxis) e.g., rhizobia and legume, and other beneficial organisms as plant-growth promoter bacteria and fungi [94–96]. With all the diversity of plant roots, exudates, microbes, and other soil organisms it is expected that in soil occur different interactions and responses which will be reflected in the soil food web structure, diversity, and function [87].

Complexity of soil food web involves the species number and the number of different kind of species (trophic and functional groups); other characteristics as connectedness, interactions strength and length of chains are important in the food web stability [97]. Throughout the soil food web the main relationship is the vital need to obtain food and energy to accomplish the life functions and the species survival; however this relationship acquire different tonalities when each microbe or protozoa; or arthropod species in the soil, develops different life strategies in response to specific natural selection forces, and they establish inter- and intraspecific biological interactions like mutualism, competition, parasitism, predation, pathogenism, fungistasis, antibiosis, allelopathy, herbivory. These 'complementary' interactions among species are of ecological relevance because throughout them is built up the structure, function and diversity of the soil community; Wardle [67], talks about some of these kind of interactions as 'interference interactions' and he indicates that apply primarily to interactions among fungi, among bacteria, and between bacteria and fungi.

Microbial symbionts (endosymbionts) play vital roles inside plants, fungi, nematodes, protozoa, insects (termites, ants), etc. [45, 98–100]; they are interactions inside interactions and are *"complementary interactions"* that biological species have been developed and evolved throughout time and space, improving their fitness [25, 45, 56]; without them the host species would be unable to live. All kind of symbionts are also components in the soil foo web; some of these relationships are obligate or facultative, and others are intermediate between an obligate and facultative behavior, depending on the press of natural selection forces and the evolutionary time through which these species have been related [101]. What about the role pathogens play in the soil food-web, and the role plant pathogens have in different important functions for plant life: seed germination [77], seedlings and young plant establishment [76], plant sexual reproduction and sexuality expression [102, 103] and their role in plant community successional process? Ecological functions of pathogens and specifically of plant pathogens, have received few attention in the plant communities of wild ecosystems [104–106], even though Dinoor and Eshed [105] drew the attention about the importance of studying plant pathogens and disease they cause in natural ecosystems, to better understand plant diseases in agroecosystems and applied the best management strategies.

To this point, we have seen that soil organisms play multiple and different activities in the soil ecosystem and all of them are relevant for the soil dynamic functioning (**Table 1**); in fact, the FAO Report of 2020 talks about "The Multifunctionality of Soil Biodiversity" [107]. But what drives soil microbial diversity? Soil ecologists suggest that innate soil spatial heterogeneity, or patchiness, would be a main environmental factor to explain soil biodiversity at different spatial and temporal scales, arguing that soil heterogeneity *'provides unrivaled potential for niche partitioning, or resource and habitat specialization, leading to avoidance of competition and hence coexistence of species'* [1]. At the same time, the knowledge of the multiple and diverse biological intra- and interspecies interactions that happen in the soil environment at all levels of biological organization, and taking as a fundamental basis the structure and complexity of the soil food web, we reason that these two factors: soil spatial heterogeneity and ecological interactions (trophic and complementary interactions) working together through evolutionary processes and time, at the population and community levels, have resulted in the immense soil biodiversity.

Pathogens affect all groups of organisms: plants, microbes, protozoa, nematodes, insects, etc.

Food webs, and therefore soil food webs, are biological systems organized with different subsystems (trophic levels), and sub-subsystems (functional groups in each trophic level). Food webs are also open systems, with a spatio-temporal dynamic of the whole, in which each subsystem and each subsystem component has also its own dynamic function but interrelated with other components in the web-system. Food webs are open systems with energy and material flow, and in some cases there is also a flow of genetic information (e.g., fungi endophytes maybe transmitted through vertical or horizontal gene transfer, [47]). In this sense, ecologists have mentioned that *"The analysis of energy and material flow is considered to be fundamental to understanding the patterns and dynamics in ecosystems and the way ecosystems are organized."* [97].

Trophic level		Producers	Consumers (first level)		Predators (first level)		Predators (second level)		Higher Predators	
Biological Organization Level	Activity	Plant Roots	Bacteria	Fungi	Protozoa	Nematodes	Micro Arthropods	Earthworms	Insects, Birds, Small Mammals	Ecosystem Services performed by Soil Biodiversity at the Community Level
INDIVIDUAL	Root Biomass Productivity and release of Root Exudates	✓								
	Organic Matter Decomposition		✓	✓						
	Mineralization (N, P, C)		✓	✓						
	Nutrient Cycling		✓	✓	✓	✓				
	Induction of Plant Resistance (ISR, AR)		✓	✓	✓	✓				Water Reservoir ↓
	Signaling Pathway in Plant Growth Promotion		✓	✓	✓	✓				Weather Buffering ↓
	Development of toxins against herbivores			✓						Soil Health
	Abiotic Stress Tolerance		✓	✓						
	Enhancement of Plant Fitness		✓	✓						↓ Healthy Plants ↓
POPULATION	Regulation of Population Densities		✓	✓	✓	✓	✓	✓	✓	
	Regulation of Populations Dispersal and Spatial Pattern		✓	✓	✓	✓	✓	✓		Oxigen Release ↓
COMMUNITY	Enhancement of Physical and Chemical Soil Environment (soil structure, aeration and water movement and holding capacity)					✓		✓	✓	Soil Conservation ↓
	Influence in the dispersal and spatial pattern of plant species in a community	✓	✓	✓	✓	✓	✓	✓	✓	Climate Regulation ↓

Pathogens affect all groups of organisms: plants, microbes, protozoa, nematodes, insects, birds, mammals, etc., in different life-strategies and processes, e.g., seed germination, seedling establishment, plant sexual reproduction, sexuality expression, successional process, impaired competition.

✓ = *Indicates participation in the activity*

Table 1.
Multiple and different functions performed by the soil organisms at the individual, population and community organization levels in the soil ecosystem and soil food web, considering the ecosystem services.

With all the information presented here, it is argued that most of the trophic relationships in the soil trophic web has been established through evolutionary processes, and this is an important basis to understand that disturbances (e.g., plant disease epidemics) in the soil food web can have irreparable consequences in the ecosystem functions, e.g., natural or managed systems. Disturbances that occur in the soil environment will cause changes at different physical and biological levels; these perturbations will affect the ecological interactions and depending on the strength and duration of the perturbation, soil ecosystem will be able to recover through its resilience and resistance capacities, expressed at the individual (e.g., dormancy), population (temporal and spatial population density and dynamics) and community (regulation throughout mutualistic vs. antagonistic relations) levels [108]. These processes will be evidenced in the complexity of the structure and diversity [108, 109] and in the stability of the soil food web [97]; However, we must also keep in mind that all the multiple and different soil organisms since plant roots to microorganisms and to animals, all together as a whole, participate in vital soil processes such as biogeochemical and nutrients cycling, soil formation and conservation, and climate regulation [2] (**Table 1**). In this sense De

Ruiter and Moore [110], indicated that *'soil food webs are thought to govern major components in the global cycling of materials, energy and nutrients'.*

Soil ecologists indicate that soil food web complexity improves the turnover of nutrients, enhance soil structure, water holding capacity and infiltration, promotes disease suppressiveness, pollutants degradation and biodiversity [18, 110]. All the different interactions that soil fungi and bacteria have established with other soil microorganisms like protozoa, nematodes, rotifers, microarthropods, mites, ants, and root plants, establish the foundations for a complex soil food web with direct and indirect biological and ecological relationships. Complexity and performance of the soil food web is also affected by physical and chemical soil factors; soil structure, particles aggregation, pore size [111], soil texture, pH [22], amount of organic matter, all of them affect direct and indirectly the soil biological species diversity, their interactions, their population densities and their spatio-temporal dynamics. At the same time, every biological activity performed by the soil community will transcend and affect some physical and chemical characteristics in soil. It is important to mention that trophic and non-trophic relationships in the soil community and ecosystems have been developed through time and they are ruled by natural selection forces, which mean, trophic food webs have evolutionary and ecological basis [101].

5. Root pathogens: soil borne plant pathogens

Soil Borne Plant Pathogens include organisms from bacteria, fungi, oomycetes, nematode, protozoa groups and mesobiotic entities like virus and viroids [11, 73, 112]. SBPP penetrate, infect and invade the roots using different biochemical and physical mechanisms, causing cell and tissue damages; they feed and establish at different regions in the roots including xylem and phloem, but they also obtain their food and energy from rhizosphere root exudates. SBPP are endo-, ecto-, or semi-endo-pathogens what means they can enter and live their whole life cycle inside roots, or they live some life stages in the soil. In plant pathogenic nematodes, there are species in which young and immature females penetrate only half of their body into the epidermal and cortex cells in the roots (*Tylenchulus semipenetrans*), until they mature and transform into a swollen body containing eggs that are released into the soil (*Meloidogyne* sp.); some plant nematodes are sedentary (*Xiphinema* sp.) while others migrate inside the roots (*Pratylenchus* sp., *Radophulus* sp.) or go up to the stem (*Ditylenchus* sp.) [11, 113]. Soil borne plant pathogens produce localized or systemic damages; they damage the roots producing root rots, wilts, necrosis and death [114] which impair nutrients and water uptake to the upper plant organs, where damage is manifested as seedlings damping-off, stunt, chlorosis, wilts, bark cracking, twigs and branch diebacks, drop of flowers and fruits, and in consequence biological and economical yield losses [12, 112, 115–117]. SBPP are obligate or facultative pathogens; some of these pathogens may live also as soil saprophytic organisms depending on substrate availability and soil environmental conditions. When soil environment conditions are adverse, many SBPP develop resistance structures (e.g., sclerotia, cysts, oospores, chlamydospores), and they enter in a dormancy stage for until 20 (*Sclerotium cepivorum*) or 30 years (*Rhizoctonia solani*) [11, 73, 118].

SBPP produce different kinds of propagules, which refers to any entity or unit able to multiple, disperse and conserve the pathogen population, e.g., in fungi: spores, conidia, sporangia, sclerotia; in nematode: eggs, cysts, immature stages, adult male and female; in virus: they are vectored by fungi, nematodes and mites [11]. Propagules and resistance structures of SBPP remain in soil as inoculum that is dispersed by water, microfauna, and cultural practices. There are important and unique pathogen traits which contribute to their successful establishment, increase and eventual epidemic expression, and these are inoculum density, pathogenicity, virulence, dispersal ability, reproductive mode (sexual/asexual), secondary hosts, and long-term resistance and survival structures in soils [12, 119]. SBPP are another natural component of the soil communities and part of the soil food web, they interact with plant roots, but they also interact with other soil organisms in mutualist and antagonist relationships [78, 120], and all these interactions shape the SBPP population density and dynamics in time and space. Some examples of important SBPP affecting plant communities and crops are, fungi: *Fusarium oxysporum*, *Rhizoctonia solani*, *Verticillium dahliae*, *Armillaria mellea*, *Gaeumannomyces graminis*, *Sclerotium ceviporum*; Oomycetes: *Phytophthora cinnamomi*, *P. capsici*, *Pythium aphanidermatum*; nematodes: *Meloidogyne incognita*, *Nacobbus aberrans*; bacteria: *Ralstonia solanacearum*, *Agrobacterium tumefaciens*; mesobiotic entities: lettuce necrotic stunt virus (LNSV), spindle tuber of potato viroid. Economic and yield crop losses caused by SBPP diseases are significant and may provoke loss of the total crop yield as in the white root rot of onion and garlic caused by the fungus *Sclerotium cepivorum* when inoculum density is high and persists in soil for long time [117].

5.1 Function of root pathogens

Plant (Root) pathogens has also been considered as microherbivores [91], which in the process of feeding from plants they release different enzymes (cellulases, chitinases) and develop different structures (fungi: haustorium, appressorium) or used structures like the nematodes stylet to enter de (root) plant tissues; at the same time, they also elicit defense/resistance mechanisms by the host plant. Plant pathogens are important components in different ecological processes of the plant community like structure and succession, development (expression) of sexuality, seed germination and establishment of seedlings [76, 105, 121–123], competition between plant species [77] and expression of allelopathy [124]. Since an evolutionary point of view, plant pathogens are important drivers of evolution of both species, the host plant and pathogen [13, 106], and therefore in the diversity of the two species involved in the process of pathogenicity [78, 125, 126].

5.2 Diseases caused by root pathogens in wild ecosystems

We know SBPP are natural components in wild ecosystems and they participate in important ecological and evolutionary processes in plant communities; however they may also cause severe and destructive epidemic diseases in nature system, as it happens in Jarrah (*Eucalyptus marginata*) forests in Western Australia which have been devastated during the last 100 years by the dieback disease epidemics caused by the Oomycete *Phytophthora cinnamomi* Rands, an exotic root pathogen introduced into Australia in the XIX century, with a host range over than 2000 plant species in more than 48 botanic families [127–129]. The destructive effect of this SBPP has caused cascade negative effects on the Australian forest ecosystem because affects indirectly, different species of insects, birds and small mammals who use to feed on the plant species destroyed by *P. cinnamomi* [130]. Dieback disease epidemic have destroyed large areas of the jarrah forest to the point that they are known as black gravel or graveyard sites because these sites are devoid of the plants and animals they supported [131]; this epidemic disease has disrupted the aboveground food web, and certainly the belowground soil food web. Plant disease epidemics caused by SBPP in wild ecosystems are uncommon so far, but they may be quite destructive, threatening entire plant communities and ecosystems [130].

6. Resilience in soils

Belowground roots, exudates, microbial and microorganisms diversity conform a complex and diverse soil food web with multiple trophic and complementary relationships, which can be classified as mutualists (+) and antagonists (−) relationships that in theory must result in a well-balanced soil system where is expressed the suppressiveness to SBPP; these are call "suppressive soils". In the development of the biological control of SBPP the concept of suppressive soils has been a key one, because take in consideration that in the soil ecosystem there are a whole microbial (fungi and bacteria) community with the potential to interact with root pathogens and regulate their population densities and dynamic under certain physical and chemical soil environmental conditions. Baker and Cook [21] originally defined suppressive soils as "*soils in which the pathogen is not able to establish or persist, the pathogen establishes but causes no damage, or the pathogen causes some damage, but the disease becomes progressively less severe, even though the pathogen persists in soil*". Development of molecular biology methodologies has allowed to better understand that in the soil microbial community, certain groups of bacteria and fungi are involved in soil suppressiveness. Mendes et al. [132] found thousands of bacteria and archaeal species in the groups of Proteobacteria, Firmicutes and Actinobacteria constantly associated with suppression to disease caused by Rhizoctonia solani in beet. In other research Penton et al. [133], resolving soil disease suppression to *R. solani* strain AG8 and *Fusarium psudograminearum* in cereal fields in Australia, found that suppressive soils were attributed to less than 40 genera of fungi, including certain endophytic species and mycoparasites in the groups Ascomycota, Basidiomycota and Chytridiomycota; the fungi genera most associated with the suppressive fields were *Xyllaria* (endophyte), *Bionectria* (mycoparasite), *Anthostomella* (saprotrophic), and also with antifungal activity *Chaetonium*, *Corynascus* and *Microdiplodia*; these authors indicate the importance of analyzing soil suppressiveness including both fungi and bacteria, and their interactions with fungi and plants. It has been mentioned that ecosystems as open systems, have the ability of buffering negative stresses throughout their resilience and resistance properties expressed at the population, community, and ecosystem levels [108], but this ability depends on the strength and time of duration that the stress factor persists. Suppressiveness of soils to SBPP and diseases may be considered as an expression of the soil resilience capacity where soil microorganisms multifunctionality must play an important role. Soil resilience is defined as the ability of a soil to recover to its initial state after a stress event [134]. Suppressiveness/resilience may not be necessarily manifested in every soil in natural environments and even less in agroecosystems.

7. Agroecosystems and biodiversity

Agroecosystems are modified ecosystems by man in which plants, animals and microorganisms biodiversity has been altered, and sometimes decreased to a minimum number of species [135, 136]. Agroecosystems are simplified systems at different levels of their structure and function where biological interactions and relationships have been disrupted, and these disturbances are expressed at different levels of organization [136]. There is a great diversity of agroecosystems in the world, from traditional multiple cropping systems under subsistence agriculture established mainly in tropical regions [8] to highly technified and extensive monocropping systems established under intensive agriculture mostly in the temperate regions [137]. Therefore, it is expected that the soil ecosystem biodiversity, structure, and function be altered at different levels. It is important to consider that the change from a natural ecosystem into an agroecosystem will always bring alterations above- and below-ground regardless of if this is a traditional, or multiple or intensive cropping; most of the different agronomic activities applied (tillage, fallow, herbicides, manure, etc.) in a crop land will certainly cause alterations in the soil ecosystem. But how important will be these changes? Which species and functions will be affected and how this will be manifested in the soil food web structure and functions? Certainly, these are no easy questions to answer. However, if we think about the most important functions of the soil ecosystem performed by the soil biota: organic matter decomposition, nutrient release and cycling, and energy flow (soil food web and *complementary relationships*), we may decide which elements and factors to weigh for a better agroecosystem design and management, affecting as little as possible soil biodiversity and soil food web functions.

The specific agronomic requirements for the crop of interest must also be considered. Intensive and extensive agroecosystems (e.g., cereal crops), are highly uniform in their genetic, physiological, and morphological structure and function [135, 136]; and the different agricultural practices such as tillage, herbicides, sowing, improved seed, fertilizers, and pesticides are usually applied with machinery, which may cause a great disturbance in the soil system. In the other hand, in traditional multiple cropping systems there are crop plants diversity and sometimes also weed diversity, they may resemble more to a nature plant community and therefore soil alterations are expected to be less [8, 135, 136]. Several researches have documented how changes aboveground and belowground affect soil biodiversity, structure and function in agroecosystems; Wardle [138] found there were disturbances on detritus food webs because of applying different tillage (no-tillage, conventional tillage) and weed management practices. Tsiafouli et al. [139] sampled soils from different agronomic management (perennial, intensive and non-intensive) in several European countries with the objective to find out how agricultural intensifications affect soil biodiversity; they found that intensification reduced richness and diversity Shannon index of faunal groups but also the average taxonomic distinctness and average breadth of related species, this mean, agricultural intensifications causes a loss of taxonomic diversity, and in turn, soil functioning maybe affected too.

Plant roots (and some death plant residues) are, as producers, in the first soil trophic level and they are source of energy for the upper trophic levels. Some ecological studies propose productivity as key component for the structure, diversity and stability in a food web [110]. Therefore, when a natural ecosystem is changed to an agroecosystem, we expect a cascade effect that will cause a disruption in plant diversity, plant species abundance and plant community composition and in turn there will be alterations in plant productivity (including root productivity), and this will affect the soil food web diversity, complexity and stability [139].

8. Root pathogens in agroecosystems

Plant diseases and epidemics caused by SBPP in agroecosystems are common, and they cause great economical and crop yield losses [12, 140]; some of them have been well studied and documented e.g., avocado root rot caused by *Phytophthora cinnamomi* [141], *Fusarium oxysporum* wilt diseases in vegetables [142], *Ralstonia solanacearum*, a bacteria, causing vascular wilt diseases in more than 200 host plants [143]. It is important to mention that for a disease to develop there must be a susceptible plant host, a virulent plant pathogen and an environment suited to its growth, these are the three components of the conceptual model "disease triangle", a key in

Plant Pathology [144]. Plant disease epidemics and therefore soilborne disease outbreaks may occur for an increase in the pathogen populations (inoculum density),
by an increase in the host plant susceptibility (age, phenological stage, nutritional
status, genetic bakcground) or by certain enviromental conditions (biotic and
abiotic) conducive to the disease expression, e.g., biologically impoverished soil,
deficient water drainage [12, 140]. The degree of root damage or severity is generally
related to the number (inoculum density) and type of pathogens, which results in
root rot wilting, necrosis, poor growth, and stunted development (deformations),
these root alterations impair nutrient and water uptake, affecting in turn development of the whole plant with a decrease in biological and economical crop yield
and food quality [114, 117]. Increments in SBPP populations and damages in roots
affect development and kind of exudates, which may affect other microbial populations altering in turn soil food webs, in a cascade effect. Management of SBPP and
disease outbreak in crops, consider the addition of organic matter, cover crops, green
manures, composts, crop rotation, and multicropping as adequate strategies because
they decrease pathogen soil inoculum through enhancement of antagonistic relationships, they also improve plant growth and resistance, multiplication of beneficial soil
microorganisms, and these strategies also enhance the soil suppressiveness/resilience
[7, 145, 146].

9. Management of root pathogens and diseases they cause in agroecosystems

Since an ecological point of view, plant disease epidemics caused by SBPP in
agroecosystems are relevant because the different control strategies applied are
mainly directed to the soil, trying to decrease pathogen inoculum density and
population dynamics [116]. Among these strategies we found: 1) Application of
chemical pesticides [147]: fungicides, nematicides, antibiotics, including biocides
like Methyl Bromide, which have an evident negative effect to the soil biodiversity.
2) Biological control [148], which involve the introduction in the soil or substrate
where plants are growing, specific bacteria or fungi species or strains that function as pathogen antagonists or plant-growth promoters or drivers of the plant
host resistance; the effects of this strategy on the soil community function and
structure, are not very well known. 3) Plant host resistance [149, 150], obtained
through the genetic improvement of the crop species by breeders using traditional
breeding techniques or modern genetic engineering methodologies introducing
resistance genes into the host crop (genetically modified organisms GMO); the
effects of this modified organisms in the soil community and soil food web, are
also not well known. 4) Cultural Management [7, 151], like soil quarantine, soil
disinfestation, intercropping and crop rotation, tillage, planting date and plant
spacing. 5) Management of the soil environment such as mulching, biofumiga-
tion, composts, and composts added with pathogen antagonists [7]. Management
strategies 4 and 5 have a direct effect and alterations on the soil communities
and food web, all these strategies imply application of organic matter into the
soil, However, several questions surge: What is the best way to apply it? Is it the
right kind of organic matter for a specific SBPP and disease? How long last their
effects? How important is the application of organic matter and the introduction
of antagonists in the soil community? What groups or species of antagonistic
organisms are the most adequate? How will these strategies affect the soi microbial
community? How the soil food web will be disrupted? How much these methods
affect important soil functions? Until now, there is not enough ecological research
about how much soil biodiversity is affected, mainly microbiome diversity, and

therefore soil food web structure and function, as a result of the strategies applied for root pathogens management. Here is interesting to mention work by Wolfgang et al. [152], with tomato crop in Uganda, where they screened from rhizosphere and surrounding soil the different groups of bacteria antagonistic to the root-knot nematode *Meloidogyne* spp. and several root pathogenic fungi, e.g., *Fusarium oxysporum*, *Botrytis cinerea*, *Sclerotium rolfsii* and *Verticillium dahliae*. *Meloidogyne* spp. is a SBPP involved in several complex diseases interacting with other plant pathogens and damaging important vegetable crops like tomato around the world and causing great crop losses. Researchers find out that infection with nematodes was correlated with a strong bacterial community shift in tomato roots, with a microbiome from healthy plants differing from infected roots, and they concluded that the different functions performed by the antagonistic microbes, including volatile organic compounds, all together can lead to synergistic beneficial effects preserving the stability and diversity of macro- and microhabitats. Their results show that rhizosphere and surrounding soil microbes, function in a complementary conjunction, performing multiple roles in a complex and dynamic system. Must be said, that there is need of research on these topics in agroecosystems with a wider view, taking into consideration soil ecological processes and principles.

10. Proposals for SBPP and diseases management, under ecological and sustainable soil biodiversity and conservation

The study and management of plant diseases is now supported for epidemiological concepts and methodologies, which allow to understand the diseases as dynamic spatio-temporal processes at the population and community levels, involving tree main components: the pathogen, the host plant and the environment (biotic and abiotic), forming the 'disease triangle', a key concept in the study of pathosystems [140, 144]. In this way, plant disease epidemiology, specifically temporal and spatial quantitative analysis of the pathosystem, set up the basis for the design of better disease management strategies in both airborne and soilborne pathosystems [12, 140]. However, this epidemiological approach must be enhanced applying ecological concepts, principles and methodologies that enrich and preserve soil diversity and the soil food web structure and functions, applying organic agriculture, composting, crop rotations and green manure [7]. A relevant consideration is to perform epidemiological research in soilborne plant diseases in long-term studies (5–7 years), under a regional (landscape) level [153] with different genetic populations of the pathogen and the host, which allow to find out how their populations are structured and have coevolved, adding another important element for understanding the genetic and evolutionary basis of the diseases, and their relation with the soil microbial community and the soil food web structure and function [13, 154]. Another especially important consideration is about agricultural intensification [155], which Tsiafouli et al. [139] demonstrated that intensification has a consistent negative effect across most soil food web components and that is not limited to specific groups of soil biota; this implies the urgent need to redesign our agroecosystems in such a way to preserve soil biodiversity and the soil ecosystem. At present, an interesting proposal for the SBPP management is the genetic redesigning of beneficial and pathogenic microorganisms of plant, soil, and root rhizosphere, pursuing the development and enhancement of soil suppressiveness and plant host resistance, from the lowest biological organization level [156]. However, several questions arise: Are we taking in consideration the importance of and the transcendental evolutionary and coevolutionary relationships among the species involved in pathosystems, in the soil food webs, in the soil ecosystem? Certainly, SBPP and diseases management strategies need to be understood following and applying ecological principles, but also evolutionary principles [15, 157].

11. Conclusions

Soil ecosystem is the support for maybe every living on earth; their development and evolution takes thousand and millions of years; however, agroecosystems have caused great changes in soils worldwide, and in many cases soils have been impoverished, run out of nutrients and organic matter, contaminated and even eroded; which means soils biodiversity structure and functions have also been greatly disrupted, and in consequence vital soil functions such as organic matter decomposition and nutrient cycling and release. Plant diseases and disease epidemics caused by SBPP have relevance in soil ecology because most strategies applied for their management are directed to the soil, affecting biological, physical and chemical soil characteristics, and altering soil diversity and soil food webs. However, human societies need to produce enough and healthy food, and soil ecosystem is the source from where to obtain healthy crops; therefore, agroecosystems must be redesigned urgently, based in the knowledge of above- and below-ground communities structure and function, and diversity conservation, to develop a sustainable agriculture with minimal impact of agricultural practices on the environment and taking care of maintaining or improving soil fertility. There is need of an integral interdisciplinary research of SBPP and diseases, considering these pathogens and processes as dynamic components of the soil ecosystem, where the analysis of the soil food webs and complementary interactions be a fundamental aspect for their management, involving epidemiological, ecological, and evolutionary principles and methodology.

Author details

María del Pilar Rodríguez Guzmán
Instituto de Fitosanidad, Fitopatología, Colegio de Postgraduados, Mexico

*Address all correspondence to: pilarrg@colpos.mx

IntechOpen

References

[1] Bardgett, R. D., Yeates, G. W. and J. M. Anderson. 2007. Patterns and Determinants of Soil Biological Diversity. In: Bardgett, R. D., Usher, M. B. and D. W. Hopkins (eds.). Biological Diversity and Function in Soils. P.p.: 100-118. Cambridge University Press. Cambridge, U.K.

[2] Coleman, D. C. 2013. Soil Biota, Soil Systems and Processes. Encyclopedia of Biodiversity, Volume 6, p.p.: 580-589. http://dx.doi.org/10.1016/B978-0-12-384719-5.00128-3

[3] Neher, D. A. 1999. Soil community composition and ecosystem processes. Comparing agricultural ecosystems with natural ecosystems. Agroforestry Systems 45: 159-185.

[4] Parton, W. J., Schimel, D. S., Cole, C. V. and D. S. Ojima. 1987. Analysis of factors controlling soil organic matter levels in Great Plains grasslands. Soil Sci. Soc. Am. J. 51:1173-1179

[5] Hawes, M. C., Brigham, L. A., Wen, F., Woo, H. H. and Y. Zhu. 1998. Function of root border cells in plant health: Pioneers in the Rhizosphere. Annu. Rev. Phytopathol. 36: 311-27.

[6] Walker, T. S., Bais, H. P., Grotewold, E. and J. M. Vivanco. 2003. Root Exudation and Rhizosphere Biology. Plant Physiology, Vol. 132, pp. 44-51.

[7] Smith, J. L. and H. P. Collins. 2007. Management of Organisms and Their Processes in Soils. In: Paul, E. A. (ed.) Soil Microbiology, Ecology, and Biochemistry. P.p.: 471-502. Third Edition. American Press. Elsevier. New York. P.p.: 471-502.

[8] Francis, Ch. A. 1986. Multiple Cropping Systems. Macmillan Publishing Company. New York. 383 pages.

[9] Kazakova-Mateva, Y. and D. Radeva-Decheva. 2015. The role of agroecosystems diversity towards sustainability of agricultural systems. Paper prepared for presentation at the 147th EAAE Seminar 'CAP Impact on Economic Growth and Sustainability of Agriculture and Rural Areas', Sofia, Bulgaria, October 7-8, 2015. 16 pages.

[10] Lemaire, G., Corvalho, P., Kronberg, S. and S. Recous. 2018. Agroecosystems Diversity. Reconciling Contemporary Agriculture and Environmental Quality. Academic Press. 478 p.

[11] Agrios, G. N. 2005. Plant Pathology. 5th Edition. Academic Press. https://doi.org/10.1016/C2009-0-02037-6

[12] Campbell, C. L. and D. M. Benson. 1994. Epidemiology and Management of Root Diseases. Springer Verlag. New York. 344 pages

[13] Burdon, J. J. and P. H. Thrall. 1999. Spatial and Temporal Patterns in Coevolving Plant and Pathogen Associations. Supplement. Am. Nat. 1999. Vol. 153, pp. S15–S33. DOI: 10.1086/303209

[14] Wisz, M. S., Pottier, J. W., Kissling, D., Pellissier, L., Lenoir, J., Damgaard, C. F., Dormann, C. F., Forchhammer, M. C., Grytnes, J. A., Guisan, A., Heikkinen, R. K., Høye, T. T., Kühn, I., Luoto, M., Maiorano, L., Nilsson, M. C., Normand, S., Ockinger, E., Schmidt, N. M., Termansen, M., Timmermann, A., Wardle, D. A., Aastrup, A. and J. C. Svenning. 2013. The role of biotic interactions in shaping distributions and realised assemblages of species: implications for species distribution modelling. Biol. Rev. (2013), 88, pp. 15-30. doi: 10.1111/j.1469-185X.2012.00235.x

[15] Thrall, P. H., Oakeshott, J. G., Fitt, G., Southerton, S., Burdon, J. J., Sheppard, A., Russell, R. J., Zalucki, M., Heino, M. and R. F. Denison. 2011.

Evolution in agriculture: the application of evolutionary approaches to the management of biotic interactions in agro-ecosystems. Evolutionary Applications, 4 (2011) 200-215. doi:10.1111/j.1752-4571.2010.00179.x

[16] Bach, E. M., Ramirez, K. S., Fraser, T. D. and D. H. Wall. 2020. Soil Biodiversity Integrates Solutions for a Sustainable Future. Sustainability, 12, 2662. 20 pages. doi:10.3390/su12072662

[17] Grandy, A. S., Wieder, W. R., Wickings, K. and E. Kyker-Snowman. 2016. Beyond microbes: Are fauna the next frontier in soil biogeochemical models? Soil Biology and Biochemistry 102(5) DOI: 10.1016/j.soilbio.2016.08.008

[18] Ingham, E. 2000. Soil Biology Primer. Website: http://soils.usda.gov/sqi/concepts/soil_biology/biology.html

[19] McCully, M. E. 1999. Roots in Soil: Unearthing the Complexities of Roots and Their Rhizospheres. Annu. Rev. Plant Physiol. Plant Mol. Biol. 50:695-718.

[20] Gregory, P. J., Bengough, A. G., George, T. S. and P. D. Hallett. 2013. Rhizosphere Engineering by Plants: Quantifying Soil–Root Interactions. American Society of Agronomy. DOI:10.2134/advagricsystmodel4

[21] Baker, K., and R. J. Cook, 1974. Biological Control of Plant Pathogens. New York, NY: WH Freeman and Company, 433 pages.

[22] Ehrlich, H. L. 2006. Geomicrobiology: relative roles of bacteria and fungi as geomicrobial agents. In: G. M. Gadd (ed.). Fungi in Biogeochemical Cycles, Pp.:1-10, Published by Cambridge University Press. British Mycological Society.

[23] Dilfuza, E. 2011. Role of Microorganisms in Nitrogen Cycling in Soils. In: Miransari, M. (ed.). Soil Nutrients. Chapter 7. 18 pages. Nova Science Publishers Inc.

[24] Zhao, L., Liu, Y., Wañ '' nmng, Z., Yuan, S., Qi, J., Zhang, W., Wang, Y. and X. Li. 2020. Bacteria and fungi differentially contribute to carbon and nitrogen cycles during biological soil crust succession in arid ecosystems. Plant Soil (2020) 447:379-392. https://doi.org/10.1007/s11104-019-04391-5

[25] Bent E. 2006. Induced Systemic Resistance Mediated by Plant Growth-Promoting Rhizobacteria (PGPR) and Fungi (PGPF). In: Tuzun S. and E. Bent (eds) Multigenic and Induced Systemic Resistance in Plants. Springer, Boston, MA. https://doi.org/10.1007/0-387-23266-4_10

[26] Chunyu, L. I., Weicong, H., Bin, P., Yan, L., Saifei, Y., Yuanyuan, D., Rong, L., Xinyan, Z., Biao, S. and S. Qirong 2017. Rhizobacterium Bacillus amyloliquefaciens Strain SQRT3-Mediated Induced Systemic Resistance Controls Bacterial Wilt of Tomato. Pedosphere 27(6): 1135-1146, 2017. doi:10.1016/S1002-0160(17)60406-5

[27] Figueredo, M. S., Tonelli, M. L., Ibánez, F., Morla, F., Cerioni, G., Tordable, M. C. and A. Fabra. 2017. Induced systemic resistance and symbiotic performance of peanut plants challenged with fungal pathogens and co-inoculated with the biocontrol agent Bacillus sp. CHEP5 and Bradyrhizobium sp. SEMIA6144. Microbiological Research 197 (2017) 65-73. http://dx.doi.org/10.1016/j.micres.2017.01.002

[28] Pieterse, C. M. J., Zamioudis, C., Berendsen, F. L., Weller, D. M., Van Wees, S. C. M. and P. A. H. M. Bakker. 2014. Induced Systemic Resistance by Beneficial Microbes. Annu. Rev. Phytopathol. 2014. 52:347-75. doi 10.1146/annurev-phyto-082712-102340.

[29] Ryu, Ch. M., Farag, M. A., Hu, Ch. H., Reddy, M. S., Kloepper, J. W. and P. W. Paré, 2004. Bacterial Volatiles Induce Systemic Resistance in Arabidopsis.

Plant Physiology, March 2004, Vol. 134, pp. 1017-1026, www.plantphysiol.or

[30] Yamamoto, F., Iwanaga, F., Al-Busaidi, A. and N. Yamanaka. 2020. Roles of ethylene, jasmonic acid, and salicylic acid and their interactions in frankincense resin production in Boswellia sacra Flueck. trees. Scientific Reports. Nature Research (2020), 10:16760 | https://doi.org/10.1038/s41598-020-73993-2

[31] Shanahan, P., O'Sullivan, D. J., Simpson, P., Glennon, J. D. and F. O'Gara. 1992. Isolation of 2, 4-diacetylphloroglucinol from a fluorescent pseudomonad and investigation of physiological parameters influencing its production. Applied and Environmental Microbiology 58 (1): 353-358. DOI:10.1128/AEM.58.1.353-358.1992

[32] Koumoutsi, A., Chen, X. H., Henne, A., Liesegang, H., Hitzeroth, G., Franke, P., Vater, J. and R. Borriss. 2004. Structural and functional characterization of gene clusters directing nonribosomal synthesis of bioactive cyclic lipopeptides in Bacillus amyloliquefaciens Strain FZB42. Journal of Bacteriology, Vol. 186, No. 4, p. 1084-1096. DOI: 10.1128/JB.186.4.1084-1096.2004

[33] Meziane, H., Van der Sluis, I., Van Loon, L. C., Höfte, M. and P. A. H. M. Bakker. 2005 Determinants of Pseudomonas putida WCS358 involved in inducing systemic resistance in plants. Mol. Plant Pathol. 6 (2):177-185. DOI: 10.1111/J.1364-3703.2004.00276.X

[34] Pal, K. K. and B. McSpadden Gardener, 2006. Biological Control of Plant Pathogens. The Plant Health Instructor DOI: 10.1094/PHI-A-2006-1117-02

[35] Arshad, M. and W. T. Frankenberger, 1997. Plant Growth-Regulating Substances in the Rhizosphere: Microbial

Production and Functions. Advances in Agronomy, Vol. 62, p.p. 45-151. https://doi.org/10.1016/S0065-2113(08)60567-2

[36] Gadd, G.M. (ed). 2006. Fungi in Biogeochemical Cycles. Part of British Mycological Society Symposia. University of Dundee. United Kingdom. Cambridge University Press. British Mycological Society.

[37] Vetter, J. 1999. The role of fungi in the Nitrogen and Carbon cycles. Acta Microbiologica et Immunologica Hungarica, 46 (2-3), pp. 197-198. DOI: 10.1556/AMicr.46.1999.2-3.6

[38] NOAA. 2021. What's The Carbon Cycle? National Ocean Service Website. https://oceanservice.noaa.gob/facts/carbon cycle. 2/26/21.

[39] Thorn, R. G. and M. D. J. Lynch. 2007. Fungi and Eukaryotic Algae. In: Eldor, A. P. (ed). Soil Microbiology, Ecology, and Biochemistry. Chapter 6. Pp.: 145-162. Academic Press. Elsevier. London.

[40] Hawksworth, D. I. and G. M. Mueller. 2005. Fungal Community: Their Diversity and Distribution. In: Dighton, J., White, J. F. and P. Oudemans (eds.). The Fungal Community. Its Organization and Role in the Ecosytem. Third Edition. P.p.: 27-37. Taylor & Francis. Mycology. Vol. 23. Boca Raton, FL.

[41] Remy, W., Taylor, T. N., Hass, H. and H. Kerp. 1994. Four hundred-million-year-old vesicular arbuscular mycorrhizae. Proc. Natl. Acad. Sci. USA, Vol. 91, pp. 11841-11843.

[42] Rodriguez, R. J., White, J. F., Arnold, A. E. and R. S. Redman. 2009. Fungal endophytes: diversity and functional roles. The New Phytologist, 182(2), 314-330. https://doi.org/10.1111/j.1469-8137.2009.02773.x

[43] Hodge, A. and A. H. Fitter. 2010. Substantial nitrogen acquisition by arbuscular mycorrhizal fungi from

organic material has implications for N cycling. Proceedings of the National Academic of Science. vol. 107, no. 31, 13754-13759 | www.pnas.org/cgi/doi/10.1073/pnas.1005874107

[44] Hodge, A. and K. Storer. 2015. Arbuscular mycorrhiza and nitrogen: implications for individual plants through to ecosystems. Plant Soil 386, 1-19. DOI:10.1007/s11104-014-2162-1

[45] Jansa, J., Bukovská, P. and M. Gryndler 2013. Mycorrhizal hyphae as ecological niche for highly specialized hypersymbionts – or just soil free-riders? Frontiers in Plant Science. doi: 10.3389/fpls.2013.00134

[46] Ahmad, I., Jiménez-Gasco, M. M. and M.E. Barbercheck. 2020. The Role of Endophytic Insect-Pathogenic Fungi in Biotic Stress Management. In: Giri, B. and M. P. Sharma (eds). Plant Stress Biology. Pp.: 379-400. Springer, Singapore. https://doi.org/10.1007/978-981-15-9380-2_13.

[47] Moonjely, S., Barelli, L. and M. J. Bidochka. 2016. Insect Pathogenic Fungi as Endophytes. In: Lovett, B. and R. J. St. Leger (eds.). Genetics and Molecular Biology of Entomopathogenic Fungi. Chapter 4. p.p.: 107-135. Advances in Genetics. Vol. 94. Brock University, St. Catharine, ON, Canada. doi.org/10.1016/bs.adgen.2015.12.04

[48] Khoyratty, S., Dupont, J., Lacoste, S., Palama, T. L., Choi, Y. H., Kim, H. K., Payet, B., Grisoni, M., Fouillaud, M., Verpoorte, R. and H. Kodja. 2015. Fungal endophytes of Vanilla planifolia across Réunion Island: isolation, distribution, and biotransformation. BMC Plant Biology 15:142. DOI 10.1186/s12870-015-0522-5

[49] Batten, J. S., Scholthof, K. B. G., Lovic, B. R., Miller, M. E., and R.D. Martyn. 2000. Potential for biocontrol of monosporascus root rot/vine decline under greenhouse conditions using hypovirulent isolates of Monosporascus

cannonballus. Eur. J. Plant Pathol. 106:639-649.

[50] Alabouvette, C., Edel, V., Lemanceau, P., Olivain, C., Recorbet, G. and C. Steinberg. 2001. Diversity and Interactions Among Strains of Fusarium oxysporum: Application to Biological Control. In: Jeger, M. J. and N. J. Spence (eds.) Biotic Interactions in Plant-Pathogen Associations. P.p.: 131-158. CABI Publishing. New York.

[51] Benhamou, N., Garand, Ch. and A. Goulet. 2002. Ability of nonpathogenic Fusarium oxysporum strain Fo47 to induce resistance against Pythium ultimum infection in cucumber. Applied and Environmental Microbiology, Vol. 68, No. 8, 4044-4060. DOI: 10.1128/AEM.68.8.4044-4060.2002

[52] De Meyer, G., Bigirimana, J., Elad, Y. and M. Hofte. 1998. Induced systemic resistance in Trichoderma harzianum T39 biocontrol of Botrytis cinerea. European Journal of Plant Pathology 104: 279-286.

[53] Clay, K. 1988. Fungal endophytes of grasses: A defensive mutualism between plants and fungi. Ecology Vol. 69, No. 1, 10-16.

[54] Sornakili, A., Thankappan, S., Sridharan, A. P., Nithya, P. and S. Uthandi. 2020. Antagonistic fungal endophytes and their metabolite-mediated interactions against phytopathogens in rice. Physiol. Mol. Plant Pathol. 112, 101525.

[55] Moreno-Gavíra, A., Huertas, V., Diánez, F., Sánchez-Montesinos, B. and M. Santos. 2020. Paecilomyces and Its Importance in the Biological Control of Agricultural Pests and Diseases. Plants 9, 1746, 28 pages; doi:10.3390/plants9121746

[56] Emerson, J. B. 2019. Soil viruses: a new hope. Systems 4: e00120-19. American Society for Microbiology. Applied and Environmental Science. https://doi.org/10.1128/mSystems.00120-19.

[57] Roossinck, M. J. 2015. Plants, viruses and the environment: Ecology and mutualism. Virology 479-480 (2015) 271-277. http://dx.doi.org/10.1016/j.virol.2015.03.041

[58] Firestone, M., 2020. Soil Virus: A Rich Reservoir of Diversity. Biological and Environmental Research. University of California, Berkeley.

[59] Starr E. P., Nuccio, E. E., Pett-Ridge, J., Banfield, J. F. and M. K. Firestone. 2019. Metatranscriptomic reconstruction reveals RNA viruses with the potential to shape carbon cycling in soil. Procc. Nat. Acad. Sci. 116, 25900-25908 (2019). DOI: 10.1073/pnas.1908291116.

[60] Kimura, M., Ji, Z. J., Nakayama, N. and S. Asakawa. 2008. Ecology of viruses in soils: Past, present and future perspectives. Soil Science and Plant Nutrition 54, 1-32 doi:10.1111/j.1747-0765.2007.00197.x

[61] Boag, B. 1986. Detection, Survival and Dispersal of Soil Vectors. In: McLean, G. D., Garrett, R. G., and W. G. Ruesink, (eds.). Plant Virus Epidemics. Monitoring, Modelling and Predicting Outbreaks. P.p.: 119-145. Academic Press. New York.

[62] Roberts, A. G. 2014. Plant Viruses: Soil-Borne. In: eLS. 13 pages. John Wiley & Sons, Ltd: Chichester. DOI: 10.1002/9780470015902.a0000761.pub3

[63] Geisen, S., Koller R., Hünninghaus, M., Dumack, K., Urich, T. and M. Bonkowski. 2016. The soil food web revisited: Diverse and widespread mycophagous soil protists. Soil Biology & Biochemistry 94 (2016) 10-18. http://dx.doi.org/10.1016/j.soilbio.2015.11.010

[64] Coleman, D. C. 2008. From peds to paradoxes: Linkages between soil biota and their influences on ecological processes. Soil Biology and Biochemistry Volume 40, Issue 2, Pages 271-289.

[65] Yeates, G. W. 2010. Nematodes in Ecological Webs. https://doi.org/10.1002/9780470015902.a0021913

[66] Lubbers, I. M., van Groenigen, K. J., Fonte, S. J., Six, J., Brussaard, L. and J. W. van Groenigen. 2013. Greenhouse-gas emissions from soils increased by earthworms. NATURE CLIMATE CHANGE VOL 3, March 2013, 187-194. DOI: 10.1038/NCLIMATE1692

[67] Wardle, D. A. 2006. The influence of biotic interactions on soil biodiversity. Ecology Letters, (2006) 9: 870-886, doi: 10.1111/j.1461-0248.2006.00931.x

[68] Neher, D. A. and M. E. Barbercheck. 2019. Soil Microarthropods and Soil Health: Intersection of Decomposition and Pest Suppression in Agroecosystems. Insects Vol, 10 (12), 414, 15 pages. doi:10.3390/insects10120414

[69] Shortlidge, E. E., Carey, S. B., Payton, A. C., McDaniel, S. F., Rosenstiel, T. N. and S. M. Eppley. 2021. Microarhropod contributions to fitness variation in the common moss Ceratodon purpureus. Proc. R. Soc. B 288: 20210119. https://doi.org/10.1098/rspb.2021.0119

[70] Gupta, V. V. S. R and G. W. Yeates. 1997. Soil Microfauna as Bioindicators of Soil Health. In: Pankhurst, C., Doube, B. M. and V. V. S. R. Gupta (eds.). Biological Indicators of Soil Health. P.p.: 201- 233. CAB International. Oxford, UK.

[71] Menta, C. and S. Remelli. 2020. Review. Soil Health and Arthropods: From Complex System to Worthwhile Investigation. Insects. 11, 54, 21 pages; doi:10.3390/insects11010054

[72] Van Straalen, N. M. 1997. Community Structure of Soil Arthropods as a Bioindicator of Soil Health. In: Pankhurst, C., Doube, B. M. and V. V. S. R. Gupta (eds.). Biological Indicators of Soil Health. P.p.: 235-264. CAB International. Oxford, UK.

[73] Koike, S. T., Subbarao, K. V., Davies, R. M. and T. A. Turini. 2003. Vegetable Diseases Caused by Soil-Borne Pathogens. Publication 8099, Division of Agriculture and Natural Resources. University of California. https://anrcatalog.ucanr.edu/pdf/8099.pdf

[74] Teng, P., (ed). 1987. Crop Loss Assessment and Pest Management. St. Paul, MN. APS Press.

[75] Dobson, A. and M. Crawley. 1994. Pathogens and the structure of plant communities. Trends in Ecology & Evoluion. Volume 9, Issue 10, October 1994, Pages 393-398.

[76] Packer, A. and K. Clay. 2000. Soil pathogens and spatial patterns of seedling mortality in a temperate tree. NATURE, VOL 404, 16 MARCH 2000. DOI: 10.1038/35005072

[77] Van der Putten, W. H. and A.M. P. Bas. 1997. How Soil-Borne Pathogens May Affect Plant Competition. Ecology, Vol. 78, No. 6, pp. 1785-1795.

[78] Van der Putten, W. H. 2000. Pathogen-driven forest diversity. Plant pathology. Views and Reviews. NATURE, Vol. 404, 232-233.

[79] D'Hertefeldt, T. and W. H. van der Putten. 1998. Physiological integration of the clonal plant Carex arenaria and its response to soil-borne pathogens. OIKOS 81: 229-237.

[80] Gelsomino, A., Keijzer-Wolters, A. C., Cacco, G. and J. D. van Elsas. 1999. Assessment of bacterial community structure in soil by polymerase chain reaction and denaturing gradient gel electrophoresis. Journal of Micro-biological Methods Volume 38, Issues 1-2, pp. 1-15.

[81] Garbeva, P., Postma, J., van Veenand, J. A. and J. D. van Elsas. 2006. Effect of above-ground plant species on soil microbial community structure and its impact on suppression of Rhizoctonia solani AG3. Environmental Microbiology (2006) 8(2), 233-246 doi:10.1111/j.1462-2920.2005.00888.x

[82] Berg, G. and K. Smalla. 2009. Plant species and soil type cooperatively shape the structure and 2 function of microbial communities in the rhizosphere. FEMS Microbiol. Ecol. 68 (2009) 1-13. DOI:10.1111/j.1574-6941.2009.00654.x

[83] JRC.ESDAC. 2921. Joint Research Centre. European Soil Data Centre (Esdac). Soil Biodiversity. https://esdac.jrc.ec.europa.eu/themes/soil-biodiversity. April 14, 2021.

[84] Dennis, P. G., Miller, A. J. and P. R. Hirsch. 2010. Are root exudates more important than other sources of rhizodeposits in structuring rhizosphere bacterial communities? FEMS Microbiol Ecol 72 (2010) 313-327.

[85] McNear, D. H. 2013. The Rhizosphere - Roots, Soil and Everything In Between. Nature Education Knowledge 4(3):1. Soil, Agriculture, and Agricultural Biotechnology. https://www.nature.com/scitable/knowledge/library/the-rhizosphere-roots-soil-and-67500617/

[86] Bonkowski, M., and F. Brandt. 2002. Do soil protozoa enhance plant growth by hormonal effects? Soil Biology & Biochemistry 34 (2002) 1709-1715.

[87] Bonkowski, M. 2004. Protozoa and plant growth: the microbial loop in soil revisited. Tansley Review. New Phytologist (2004) 162 : 617-631, www.newphytologist.org

[88] Ferris, H. and M. M. Matute, 2003. Structural and functional succession in the nematode fauna of a soil food web. Appl. Soil Ecol. 23, 93-110.

[89] Yeates, G. W., Wardle, D. A. and R. N. Watson. 1999. Responses of soil nematode populations, community

structure, diversity, and temporal variability to agricultural intensification over a seven-year period.

[90] Dobson, A. P., and P. J. Hudson. 1986. Parasites, Disease and the Structure of Ecological Communities. Trends Ecol. & Evol. 1:11-15.

[91] Dobson, A., Lafferty, K. and A. Kuris. 2005. Parasites and Food Webs. In: Pascual, M. and J. A. Dunne (eds.). Ecological Networks: Linking Structure to Dynamics in Food Webs. Oxford University Press. P.p.: 119- 135.

[92] Gregory, P. J. 2006. Plant roots. Growth, activity and interaction with soils. Oxford. Blackwell Publishing. 318 pp.

[93] English, J. T. and D. J. Mitchell. 1994. Host Roots. In: Campbell, C. L and D. M. Benson (eds). Epidemiology and Management of Root Diseases. P.p.: 34-64. Springer-Verlag. Heidelberg.

[94] Bais, H. P., Park, S. W., Weir, T. L., Callaway, R. M. and J. M. Vivanco, 2004. How plants communicate using the underground information superhighway. Trends in Plant Science 9(1):26-32.

[95] Bais, H. P., Broeckling, C. D. and J. M. Vivanco. 2008. Root Exudates Modulate Plant-Microbe Interactions in the Rhizosphere. In: Petr Karlovsky (ed.). Secondary Metabolites in Soil Ecology. P.p.: 241-254. Springer-Verlag. Heidelberg.

[96] Berendsen R. L., Pieterse, C. M. J. and P. A. H. M. Bakker. 2012. The rhizosphere microbiome and plant health. Trends in Plant Science Vol. 17, No. 8, 478-486.

[97] De Ruiter, P. C., Neutel, A. M., and J. C. Moore. 1996. Energetics and Stability in Belowground Food Webs. In: Polis, G. A. and K. O. Wnemiller (eds.). Food Webs. Integration of Patterns and Dynamics. P.p.: 201-210. Chapman and Hall. New York.

[98] Adams, D. G., Bergman, B., Nierzwicki-Bauer, S. A., Duggan, P. S., Rai, A. N. and A. Schüßler. 2013. Cyanobacterial-Plant Symbioses. In: Rosenberg, E., DeLong, E. F., Lory, S., Stackebrandt, E. and F. Thompson. (eds). The Prokaryotes. Fourth Edition. P.p.:359-400. Springer, Berlin, Heidelberg. https://doi.org/10.1007/978-3-642-30194-0_17

[99] Kobayashi, D. Y. and J. A. Crouch. 2009. Bacterial/Fungal Interactions: From Pathogens to Mutualistic Endosymbionts. Annu. Rev. Phytopathol. 2009. 47:63-82. doi: 10.1146/annurev-phyto-080508-081729

[100] Sapountzis, P., de Verges, J., Rousk, K., Cilliers, M., Vorster, B. J. and M. Poulsen. 2016. Potential for Nitrogen Fixation in the Fungus-Growing Termite Symbiosis. Front. Microbiol. 7:1993. doi: 10.3389/fmicb.2016.01993

[101] Eklöf, A., Helmus, M. R., Moore, M. and S. Allesina. 2012. Relevance of evolutionary history for food web structure. Proc. R. Soc. B (2012) 279, 1588-1596. doi:10.1098/rspb.2011.2149

[102] Parker, M. A. 1987. Pathogen impact on sexual vs. asexual reproductive success in Arisaema triphyllum. Amer. J. Bot. 74(11): 1758-1763. 1987.

[103] Zemp, N., Tavares, R. and A. Widmer. 2015. Fungal Infection Induces Sex-Specific Transcriptional Changes and Alters Sexual Dimorphism in the Dioecious Plant Silene latifolia. PLoS Genet 11(10): e1005536. doi:10.1371/journal.pgen.1005536

[104] Burdon, J. J. and P. H. Thrall. 2011. What have we learned from studies of wild plant-pathogen associations?—the dynamic interplay of time, space and life-history. European Journal of Plant Pathology. DOI 10.1007/s10658-013-0265-9

[105] Dinoor, A. and N. Eshed. 1984. The role and importance of pathogens in

natural plant communities. Annu. Rev. Phytopathol. 22: 443-466.

[106] Gilbert, S. G. 2002. Evolutionary Ecology of Plant Diseases in Natural Ecosystems. Annu. Rev. Phytopathol. 40:13-43, doi: 10.1146/annurev. phyto.40.021202.110417

[107] FAO, ITPS, GSBI, CBD and EC. 2020. State of knowledge of soil biodiversity - Status, challenges and potentialities, Report 2020. Rome, FAO. https://doi.org/10.4060/cb1928en.

[108] Shade, A., Peter, H., Allison, S. D., Baho, D. L., Berga, M., Bürgmann, H., Huber, D. H. Langenheder, S., Lennon, J. T., Martiny, J. B. H., Matulich, K. L., Schmidt, T. M. and J. Handelsman. 2012. Fundamentals of microbial community. Resistance and Resilience. Front. Microbiol., 19 December 2012. Article 417, 19 pages. doi: 10.3389/fmicb.2012. 00417

[109] Delgado-Baquerizo, M., Bardgett, R.D., Vitousek, P.M., Maestre, F. T., Williams, M. A., Eldridge, D. J., Lambers, H., Neuhauser, S., Gallardo, A., García-Velázquez, L., Sala, E.O., Abades, S. R., Alfaro, F. D., Berhe, A. A, Bowker, M. A., Currier, C. M., Cutler, N. A., Hartn, S. C., Hayes, P. E., Hse, Z. Y., Kirchmair, M., Peña-Ramírez, V. M., Pérez, C. A., Reed, S. C., Santos, F., Siebe, C., Sullivan, V. W., Weber-Grullon, L., and N. Fierer. 2019. Changes in belowground biodiversity during ecosystem development. Proceedings of the National Academy of Sciences. PNAS. March 2019, 1-6 pages. www.pnas.org/cgi/doi/10.1073/pnas.1818400116

[110] De Ruiter, P. C. and J. C. Moore. 2005. Food–Web Interactions. Encyclopedia of Soils in the Environment, 2005: 59-67.

[111] Elliot, E. T., Anderson, R. V., Coleman, D. C., and C. V. Cole. 1980. Habitable pore space and microbial trophic interactions. Oikos 35, 327-335.

[112] Krupa, S. V. and Y. R. Dommergues. 1979. Ecology of Root Pathogens. Elsevier Scientific Publishing Company. New York. 281 pages. 1st. Edition. Academic Press. 478 pages.

[113] Palomares-Rius, J.E., Escobar, C., Cabrera, J., Vovlas, A. and P. Castillo. 2017. Anatomical Alterations in Plant Tissues Induced by Plant-Parasitic Nematodes. Front. Plant Sci. 8:1987. 16 p. doi: 10.3389/fpls.2017.01987

[114] González-Reyes H., Rodríguez-Guzmán M. P., Yáñez-Morales M. J. and J. A. S. Escalante-Estrada. 2020. Temporal dinamycs of vanilla (Vanilla planifolia) wilt disease associated to Fusarium spp. in three crop systems at Papantla, Mexico. Tropical and Subtropical Agroecosystems 23 (2020): #19. 13 pages.

[115] Milica, M., Rekanović, E., Hrustić, J., Grahovac, M. and B. Tanović. 2017. Methods for management of soilborne plant pathogens. Pestic. Phytomed. (Belgrade), 32(1), 9-24. DOI: https://doi.org/10.2298/PIF1701009

[116] Panth, M., Hassler, S. C. and F. Baysal-Gurel. 2020. Methods for Management of Soilborne Diseases in Crop Production. Agriculture, 10, 16; 21 p. doi:10.3390/agriculture10010016

[117] Ponce-Herrera, V., García-Espinoza, R., Rodríguez-Guzmán, M. P. and E. Zavaleta-Mejía. 2008. Temporal analysis of white rot (Sclerotium cepivorum Berk.) in onion (Allium cepa L.) under three pathogen inoculum densities. Agrociencia. VOL. 42 Núm. 1: 71-83.

[118] Baker, R. R. and W. C Snyder. 1965. Ecology of soil-borne plant pathogens – Prelude to biological control. University of California Press, Berkeley, Los Angeles. 571 pp.

[119] Horsfall, J. G. and E. B. Cowling. 1978. Pland Disease. An Advanced Treatise. Vol. II. How Disease Develops

in Populations. Academic Press. New York. 436 pages.

[120] McGonigle, T. P. and M. Hyakumachi. 2001. Feeding on Plant-pathogenic Fungi by Invertebrates: Comparison with Saprotrophic and Mycorrhizal Systems. In: Jeger, M. J. and N. J. Spence (eds.). Biotic Interactions in Plant-Pathogen Associations. CABI, 63-85

[121] Burdon, J. J. 1993. The role of parasites in plant populations and communities. Ecol. Stud. Anal. Synth. 99: 165-179.

[122] Hansen, E. M. and E. M. Gohen, 2000. Phellinus weirii and other native root pathogens as determinants of forest structure and process in wester North America. Annu. Rev. Phytopathol. 38: 515-539.

[123] Hansen, E. M. and J. K. Stone. 2005. Impacts of Plant Pathogenic Fungi on Plant Communities. In: Dighton, J., White, J. F. and P. Oudemans (eds.). The Fungal Community. Its Organization and Role in the Ecosytem. Third Edition. P.p.: 461-474. Taylor & Francis. Mycology. Vol. 23. Boca Raton, FL.

[124] Mattner, S. W. 2006. The Impact of Pathogens on Plant Interference and Allelopathy. In: Inderjit and K. G. Mukerji (eds.), Allelochemicals: Biological Control of Plant Pathogens and Diseases, 79-101, Springer. Netherlands.

[125] Hansen, E. M. 1999. Disease and diversity in forest ecosystems. Aust. Plant. Pathol. 28: 313-319.

[126] Van der Kamp, B. J. 1991. Pathogens as agents of diversity in forested landscapes. For. Chron. 67: 353-354.

[127] Broadbent, P., Baker, K. F. and Y. Waterworth. 1971. Bacteria and actinomycetes antagonistic to fungal root pathogens in Australian soils. Aust. J. Biol. Sci. 24 (5): 925-944. DOI: 10.1071/bi9710925

[128] Keen, B. and T. Vancov. 2010. Phytophthora cinnamomi suppressive soils. In: Méndez-Vilas A. (ed.) Current Research, Technology and Education Topics in Applied Microbiology and Microbial Biotechnology. p.p.: 239-250.

[129] Newhook F. J. and D. F. D. Podger. 1972. The role of Phytophthora cinnamomi in Australia and New Zealand Forests. Annual Review of Phytopathology Vol. 10: 299-326.

[130] Australian Government. 2010. A Guide to Managing and Restoring Wetlands in Western Australia. Phytophthora Dieback. Chapter 3: Managing Wetlands. 33 pages. Department of Environment and Conservation. Australian Government. 2010. https://www.dpaw.wa.gov.au/ images/documents/conservation-management/wetlands/Wetland_management_guide/phytophthora-dieback.pdf

[131] Burgess, T. and L. Twomey. 2009. Mysterious diversity – the protists (including the fungi). In: Calver, M. C., Lymbery, A., McComb, J. A. and M. Bamford. (eds.) Environmental Biology. Cambridge University Press, Port Melbourne, pp. 202-227.

[132] Mendes, R., Kruijt, M., de Bruijn, I., Dekkers, E., van der Voort, M., Scheider, J. H. M., Piceno, I. M., DeSantis, T. Z, Andersen, G. L., Bakker, P. A. H. and J. M. Raaijmakers. 2011. Deciphering the Rhizosphere Microbiome for Disease-Suppressive Bacteria. SCIENCE 332, pp.: 1097-1100.

[133] Penton, C. R, Gupta, V. V. S.R., Tiedje, J. M., Neate, S. M., Ophel-Keller, K., Gillings, M., Harvey, P., Phan, A. and D. K. Roget. 2014. Fungal Community Structure in Disease Suppressive Soils Assessed by 28S LSU Gene Sequencing. PLoS ONE 9(4): e93893. doi:10.1371/journal.pone.0093893

[134] GRDC. 2013. Managing Soil Organic Matter. A Practical Guide. Grains

Research and Development Corporation. Department of Agriculture and Food. Australian Western Government. 110 pages.

[135] Altieri, M. A. 1992. Biodiversidad, Agroecología y Manejo de Plagas. CLADES. CETAL Ediciones. 162 págs.

[136] Gliessman, S. R. 2002. Agroecología: Procesos Ecológicos en Agricultura Sostenible. Turrialba, Costa Rica. CATIE. 359 págs.

[137] Green, T. R., Kipka, H., David, O. and G. S. McMaster. 2017. Where is the USA Corn Belt, and how is it changing? Publications from USDA-ARS/UNL Faculty/1840. https://digitalcommons. unl.edu/usdaarsfacpub/1840

[138] Wardle, D. A., 1995. Impacts of disturbance on detritus food webs in agroecosystems of contrasting tillage and weed management practices. Advances in Ecological Research 26, 105-185.

[139] Tsiafouli, M. A., Thebault, E., Sgardelis, S. P., De Ruiter, P. C., Van Der Putten, W. H., Birkhofer, K., Hemerik, L., De Vries, F. T., Bardgett, R. D., Brady, M. V., Bjornlund, L., Jørgensen, H. B., Soren, Ch., D' Hertefeldt, T., Hotes, S., Hol, W. H. G., Frouz, J., Liiri, M., Mortimer, R., Setala, H., Tzanopoulos J., Uteseny, J., Pizl, V., Stary, J., Wolters, V. and K. Hedlund. 2016. Intensive agriculture reduces soil biodiversity across Europe. Global Change Biology (2014), doi: 10.1111/gcb.12752

[140] Madden, L. V., Hughes, G. and F. van den Bosch. 2007. The Study of Plant Disease Epidemics. APS Press, St. Paul, MN. 432 pages.

[141] Zentmyer, G. A. 1980. Phytophthora cinnamomi and the diseases it causes. Monograph, American Phytopathological Society. 1980 No.10.

[142] Armstrong, G. M. and J. K. Armstrong. 1981. Forma especiales and races of Fusarium oxysporum causing wilt diseases. In: Nelson, P. E., Toussoun, T. A. and R. J. Cook, (eds.). Fusarium: Diseases, Biology and Taxonomy. Pennsylvania State University Press. P.p: 391-399.

[143] Kelman, A. 1998. One hundred and one years of research on bacterial wilt. In: Prion, P.H., Allen, C. and J. Elpherstone (eds.). Bacterial Wilt Diseases: Molecular and Ecological Aspects. Springer. Heidelberg. P.p.: 1-5.

[144] Zadoks, J. C. and R. D. Schein. 1979. Epidemiology and Plant Disease Management. Oxford University Press. Oxford. 427 pages

[145] Lucas, P. 2006. Diseases Caused by Soil-Borne Pathogens. In: Cooke, B. M., Jones, D. G. and B. Kaye (eds.). The Epidemiology of Plant Diseases. 2nd edition, 373-386. Springer.

[146] Widmer, T. L., Mitkowski, N. A. and G. S. Abawi. 2002. Soil Organic Matter and Management of Plant-Parasitic Nematodes. Journal of Nematology 34(4):289-295. 2002.

[147] Morton, H. V. 1994. Chemichal Management. In: Campbell, C. L. and D. M. Benson (eds.). Epidemiology and Management of Root Diseases. P.p.: 276-292. Springer-Verlag. New York.

[148] Fravel, D. R. and C. A. Engelkes. 1994. Biological Management. In: Campbell, C. L. and D. M. Benson (eds.). Epidemiology and Management of Root Diseases. P.p.: 293-307. Springer-Verlag. New York.

[149] Robinson, R. A. 1987. Host Management in Crop Pathosystems. McMillan. New York. 263 pages.

[150] Shew, H. D. and B. B. Shew. Host Resistance. In: Campbell, C. L. and D. M. Benson (eds.). Epidemiology and Management of Root Diseases. P.p.: 244-275. Springer-Verlag. New York.

[151] Summer, D. R. 1994. Cultural Management. In: Campbell, C. L. and D. M. Benson (eds.). Epidemiology and Management of Root Diseases. P.p.: 309-333. Springer-Verlag. New York.

[152] Wolfgang, A., Taffner, J., Guimaraes, A., Coyne, R. and G. Berg. 2019. Novel strategies for soil-borne diseases: Exploiting the Microbiome and Volatile-Based Mechanisms Toward Controlling Meoidogyne-Based Disease Complexes. Front. Microbiol. 10: 1296. 15 p. doi: 10.3389/micb.2019.01296

[153] Plantegenest, M., Le May, C. and F. Fabre. 2007. Landscape epidemiology of plant diseases. J. R. Soc. Interface (2007) 4, 963-972. doi:10.1098/rsif.2007.1114

[154] Ampt, E. A., van Ruijven, J., Raaijmakers, J. M., Termorshuizen, A. J. and L. Mommer. 2019. Linking ecology and plant pathology to unravel the importance of soil-borne fungal pathogens in species-rich grasslands. Eur J Plant Pathol (2019) 154:141-156. https://doi.org/10.1007/s10658-018-1573-x

[155] Kleijn, D., Bommarco, R., Fijen, T. P. M., Garibaldi, L. A., Potts, S. G. and W. H. van der Putten. 2019. Ecological Intensification: Bridging the Gap between Science and Practice. Trends in Ecology & Evolution, February 2019, Vol. 34, No. 2 https://doi.org/10.1016/j.tree.2018.11.002

[156] Rana, K. L., Kour, D., Kaur, T., Devi, R., Yadav, A. N., Yadav, N., Dhaliwal, H. S. and A. K. Saxena. 2020. Endophytic microbes: biodiversity, plant growth-promoting mechanisms and potential applications for agricultural sustainability. Antonie van Leeuwenhoek 113: 1075-1107. https://doi.org/10.1007/s10482-020-01429-y

[157] Zhu, J., Thrall, P. H. and J. J. Burdon. 2014. Achieving sustainable plant disease management through evolutionary principles. Trends in Plant Science. 19(9):570-575. DOI: 10.1016/j.tplants.2014.04.010

Chapter 12

Exploiting the Attributes of Biocontrol Agent (*Neochetina bruchi*) as a Potential Ecosystem Engineer's

Prerna Gupta and Sadhna Tamot

Abstract

The biodiversity of lakes is continuously declining and diverse communities are being substituted by monoculture of invasive *Eichhornia crassipes,* resulting in a slew of environmental cascade effects. The ability of the *Neochetina bruchi* to self-perpetuate is a desirable aspect of biological control since it decreases the population to a reasonable level, making the approach more sustainable. *N. bruchi* is often referred to as "ecological engineers" because of the number of services it provides to the environment and enables herbicide application to be substantially reduced. Despite the presence of highly effective weevils against this weed, its effect on water hyacinth in association with the nutrients present in sites, is likely to vary with levels of disturbance caused by natural and anthropogenic factors. Understanding the aspects that determine the performance of these eco-engineers as valuable management tools will help to guide future endeavors. Our objective is to better comprehend their utility and limitations, along with critical knowledge gaps, to further enhance future applications.

Keywords: invasive species, *Eicchornia crassipes*, *Neochetina bruchi*, *Hoagland and Arnons Solution*

1. Introduction

Wetlands have characteristic aquatic plants called macrophytes that can survive in waterlogged soils. Macrophytes are hydrophytes of freshwater that can be easily seen with the naked eye and are normally found growing in or on the surface of water [1]. As they are primary producers, they form the base of food chains providing food for fingerlings, tadpoles, and other aquatic organisms [2]. The vegetation provides a habitat for invertebrates, protection against predators, and reproduction refuges to young fishes [2]. A total of 106 species of macrophytes were reported in Bhoj Wetland belonging to 87 genera and 46 families together with 14 rare species [3]. Macrophytes are also called the "kidneys of landscape" as they filter sediments and excess nutrients from water [4]. They act as nutrient sinks (uptake nutrients) as well as nutrient pumps (moving compounds from sediment to water column) thus influencing water chemistry [5]. Therefore, nutrient concentrations vary in different limnetic layers and the exchange of these nutrients depends upon the temperature, concentration of dissolved oxygen, and bacterial action [6].

199

Unethical human activities near wetlands have changed the nutrient dynamics, favoring the growth of overpopulated invasive species of macrophytes at the cost of native, species, thus losing biodiversity [7]. Influx of sewage causes overgrowth, aging, and subsequently decay of macrophytes, devouring the system from life-giving oxygen to anaerobic conditions. The biodiversity of wetlands is continuously declining and diverse communities are being substituted by monoculture of invasive species (e.g., *Eicchornia crassipes, Wolfia globosa,* and *Lemna minor*) [8]. After the death and decay of these plants, the scenario does not come to an end, nutrients are further released into the water that again supports the next crop of aquatic weeds. Thus, it is a continuous endless process that impacts the food web and thus the ecological integrity of wetlands [8]. The combined sum of these conditions causes inexplicably severe consequences when taken as a whole. Thus, there is a need to take some serious measures to get rid of these invasive species and conserve the integrity of our wetlands.

2. Chapter review

The Water Hyacinth is a South American perennial free-floating species [3]. It belongs to the family Pontederiaceae, a family of heterostylous flowering plants, which form monospecific, dense mats in lakes and wetlands [9]. The weed is present all over India and causes significant evaporation losses (1.26–9.84%). *Eicchornia* can give rise to 3000 new offspring in 50 days [10]. It is capable of doubling its area every 12–14 days during the growing season. It reduces light and dissolves oxygen, thus hampering the aquatic life and destroying the food web [11]. The consequences are devastating for those communities reliant on water bodies for water, food, sanitation, and transport [12].

The manual operation to eliminate macrophytes from the lakes requires a large, number of manpower. Though this method is widely used but it is time taking, laborious, uneconomical, and cannot be used for large size water bodies. There are also various types of mechanical devices used to remove aquatic weeds from water bodies but these devices have some limitations. Removal of weeds from the deeper zones of the lakes, is a major constraint as the aquatic weed re-grows up from their rootstocks in this method [13]. Mechanical devices are site-specific and nonselective, that is, they also cut native species with invasive species [13]. The disposal of enormous quantities of removed plant material is another problem in manual and mechanical control.

A wide range of chemical herbicides are now available for controlling the growth of aquatic weeds. Their lethal action is either by direct contact (contact herbicides) or by absorption (systemic herbicides) of the chemical from the treated part of the plant, affecting the biochemical pathways [10]. The chemical methods may be used in a case where the water quality maintenance is not the main issue because, after the treatment of chemicals, the water of the reservoir becomes unsuitable from a drinking point of view [14]. The use of herbicides to eradicate aquatic weeds is a short-term gain only. As long as the chemical effect remains, weed mortality can be observed but after a certain time, frequent use of chemicals is required to control weed growth. These chemicals are costly and after death aquatic weeds started to settle down at the bottom and reduce the depth of aquatic resources as well as release nutrients after decomposition, which supports the next weed [15, 16]. These chemicals also impact the food chain and other micro or macro-organisms of the aquatic system. 2,4-dichlorophenoxyacetic acid, Endothall, Copper sulfate, Diquat, and Glyphosate are some of the commonly used herbicides for the eradication of weeds [17].

As a result, the current scenario necessitates weed control strategies other than chemicals, and in this context, biological control is gaining prominence around the world. Considering the above-mentioned limitations of typically obsolete technologies compels the adoption of novel ways based on biological agents that are environmentally safer, friendlier, economical, and viable. The first serious measure was taken in the US in 1970's for the biological control of *E. crassipes*. This method controls the excessive growth of a pest by another organism which is naturally a predator.

This goal can be accomplished by natural bio-delegate, such as *N. bruchi*, which are chevroned Coleopterans belonging to class Insecta. Adult weevil is 4–5 mm long, having a brownish to grey tint. Abdomen is covered with fused brown tan "V"-shaped elytral marking [11]. Male is 3.5 mm in length with a thick and slightly curved snout. Female is 4.5 mm in length with a shiny tip, slender, and more curved snout [11]. Larvae damage the plant by forming tunnels through the petiole [12]. Adult weevils mainly feed preferentially on the slender upper branch of the petiole and the epidermis of the leaves, producing squarish feeding scars. The self-perpetuating existence of the control agents is a desirable feature of biological management, which decreases the population to a reasonable level, making the approach more sustainable [18]. The natural enemies, or control agents, have no undesirable side effects and insignificant impacts on nontarget animals or plants [15]. This method is cost-effective, environmentally acceptable, does not pollute the environment, and enables herbicide application to be substantially reduced [19]. Some of the noteworthy contribution in this field were carried out by Center [20], Center [21], Firehun [22], Heard and Winterton [23], Julien [12], Kumar [24].

The control of invasive species depends on a combination of various factors, such as temperature, nutrient level of the weed, climate and hydrology of the catchment, and number of healthy insects released [4]. The control of *E. crassipes* through biological control agents will be easier under lower nutrient conditions because plant biomass accumulation will be lower [23]. Understanding the combination of multiple drivers of plant growth, under changing conditions is essential for controlling the growth of water hyacinth [25]. Ray [18] studied the minimum required inoculation load of weevils for the control of three growth stages of water hyacinth based on fresh biomass, plant height and number of leaves and concluded that the smaller growth stages can be controlled earlier than the larger ones.

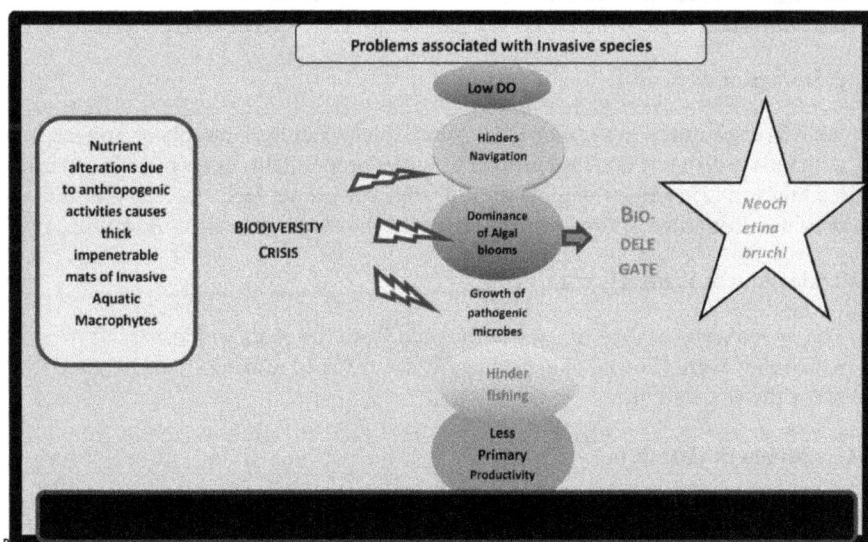

Figure 1.
Flow chart on problems associated with excess growth of Invasive species.

Based on the findings of previous studies, we should concentrate our efforts on ensuring that the control meets certain criteria, such as being effective against the target organism in a variety of nutrient conditions present in sites, which are likely to vary with levels of disturbance caused by natural and anthropogenic factors and achieving adequate control levels in the field with varying stocking densities. If such faults are not addressed, they create favorable conditions for *Eichhornia* to reproduce quicker than the weevil's growth rate. Hence, when deliberate introductions of bioagents are to be made, it is essential to release weevils first in the laboratory and assess their effects before releasing them in open field trials. In this chapter, we presented the results of a laboratory investigation to manage *E. crassipes* with *N. bruchi* (as the potential eco-engineer) in diverse nutritional situations. Thus, the study will endorse our perception of the degree to which weeds are eradicated in diverse nutritional situations, which will indirectly help to improve water quality and conserve aquatic life. Moreover, these efforts will aid in endeavoring the future application of bio-agents with the explicit goal of restoring biodiversity. The chapter comprehends the utilities and limitations, along with reviews that maintain and enhance the action of natural enemies to boost future applications (**Figure 1**).

3. Material and methods

This study was conducted in the Animal House Facility of the Department of Bioscience, Barkatullah University, Bhopal, to support studies on weed control in laboratory conditions. Stock cultures of *E. crassipes* and *N. bruchi* were obtained from Behata village behind the Sadhu Vaswani College, Bairagarh, Madhya Pradesh.

1. Collection of weevils

To collect large number of adult weevils without damaging the plants, the plants were sink under the surface of the water, by putting a rigid grid over the top of the plants. Weighing the grid down with bricks and leaving it like this for 2 hours helped in collecting the weevils from the surface of the tank with a small sieve. Authentic collection of adult pairs was a very important issue for this experiment but adults laid eggs, which was very fruitful for the experiment.

2. Storage of weevils

Weevils were stored in a plastic bucket with freshwater hyacinth plants and covered with mosquito net cloth. This process provides proper light, oxygen while ensuring feeding and restricting the flight of *N. bruchi*. The proper feed material was also provided from time to time, which supported the ovipositioning of *N. bruchi*.

3. Cleaning of Water Hyacinth

The weeds were washed to remove the mud from the roots and leaves, after which they were allowed to dry on paddy straw for 10 min and then weighed to know the wet weight.

3.1 Experimental design

The experiment was conducted in the laboratory conditions in plastic basins or tubs of size 41 cm in diameter and height 13 cm. The monitoring period was 75 days and water hyacinths were kept in 12 different plastic tubs with 10 adult pairs of *N. bruchi* in each tub. Biological agents were introduced in all the groups except their control. In these groups, different weighable 7-cm long, four water hyacinths were introduced at the primary stage. The weight of the plant was measured by the Analytical and Precision Weighing balance (Endel-JA203P). The bioagents were introduced after 7 days of system establishment.

Group T1 = Control I + Natural conditions were maintained + *N. bruchi*
Group T2 = Control II + Hoagland and Arnon solution + *N. bruchi*
Group T3 = Control III + N and P deficient Hoagland and Arnon solution, + *N. bruchi*

Generally, natural condition was maintained in T1 for up-gradation of the weed in this condition. Water was added to the T1 set to provide nutrition to the hyacinths and prevent them from drying. The fresh nutrient solution was prepared and added weekly in T2 and T3 groups. Basically, fortnight data were collected for general observation purposes. In the three groups (T1, T2, and T3) different conditions were maintained that provide different sets of data. In all the treatments, all other insects were removed immediately to maintain the original herbivore densities, and any dead adult weevils were replaced with other weevils.

3.2 Statistical analysis

The data collected were subjected to statistical analysis (one-way ANOVA) by using Excel–Mac operating system software. All the groups, with overall significance, were further compared for intergroup variation by Tukey's honestly significant differences (Tukey HSD) test.

3.3 Life cycle of *N. bruchi*

- **Eggs.** Eggs are ovoid about ¾ mm in length and change their color to pale orange as the time of hatching approaches.

- **Larval formation.** The larval stage of the *N. bruchi* starts after 1 week of laying of eggs. Larvae are white or cream-colored and the larval stage continues up to 27 days.

- **Pupa formation.** From larva to pupa formation *N. bruchi* took 29 days. The pupa of *N. bruchi* is normally found beneath the root of the water hyacinth plant. They cut off the small lateral rootlets and form a spherical parchment-like cocoon around themselves. The cocoon is formed among the lateral rootlets and attached to the main root axis below the water surface.

- **Adult formation.** *N. bruchi* is confirmed as an adult weevil after 29 days of pupa stage. They lived in the adult stage for up to 120 days.

Assessment of the effectiveness of the eco-engineer as management tools

4. Results

In the present study, the insect treatment significantly reduced the plant weight in T1 and T3 (N and P were absent) groups. However, the decrease in weed weight in the T2 (weevil exposed condition with Hoagland's solution) group was significantly lower compared to the (T1) and (T3) groups after 75 days of the growth period, as shown in **Table 1**. All herbivory treatments showed lower values of plant weight than their Controls. The change in fresh weight in all herbivory treatments and the control

Macrophytes weight (g)	Tub	March		April		May
		15th day	30th day	45th day	60th day	75th day
T1 Natural condition	A	277	335	298	265	180
	B	272	321	284	234	173
	C	281	339	268	280	168
	Control	244	260	280	330	339
T2 Hoagland and Arnon Sol	D	391	423	478	443	378
	E	388	440	467	430	369
	F	396	435	471	430	385
	Control	390	400	430	450	505
T3 N and P	G	298	350	388	322	273
	H	287	322	340	288	254
	I	292	334	330	298	268
	Control	360	355	330	323	305

Table 1.
Weight of E. crassipes reduced during the experimental period of 75 days.

was different, varying from 3.3% to 40.21% during the week 1 to 10th, as shown in **Table 2**. There were no significant differences between the treatments of plant height parameters, as shown in **Table 3**.

The results of this study showed that insect herbivory retarded the biomass and growth of *E. crassipes* and these agents were more effective in the T1 and T3 groups. The natural growth of water hyacinth was higher in March in group T1 because the bioagent (*N. bruchi*) was residing in this system in an immobile condition. In April and May, the bioagents were found in an active condition, so the wet weight of the plant was reduced in comparison to March. The weight of water hyacinth in tub C was higher than the other two groups in the second half of April month. This incident occurred because at that time population of *N. bruchi* entered the nether point, that is, they were found dwelling beneath the surface of the plant. This is why the biomass of the water hyacinth increased from 268 gm to 280 g in the second half of April month.

We are speculating that the reason for the success of this experiment in the T1 and T3 groups is that adult weevils scar the leaves of the weed, which lowers the photosynthetic rate and surface area while enabling the access and transfer of saprophytes into the leaves. The introduction of microbes via feeding scars elicits symptoms, such as increased respiration rates, poor chlorophyll and yellowing of leaves decreased buoyancy, and water and nutrient deficiencies [26–28]. All of these symptoms contribute to decreased growth and cause necrosis of plant tissue thus disrupting the plant leaf dynamics [20, 29]. Naturally, nitrogen and phosphorous

Control (%)	March (%)	April (%)	May (%)
Control I 28.023	A 35.018	B 36.397	C 40.213
Control II 22.772	D 3.324	E 4.896	F 2.77
Control III 15.277	G 8.389	H 11.498	I 8.219

Table 2.
Percentage of E. crassipes reduced by the N. bruchi during the three different nutrient conditions.

Macrophytes	Aqua rium	March		April		May
		15th day	30th day	45th day	60th day	75th day
T1 Natural condition	A	7	7.4	7.4	7.4	7.4
	B	7	7.3	7.3	7.3	7.3
	C	7	7.4	7.4	7.4	7.4
	Control	7	7.3	7.4	7.4	7.5
T2 Hoagland and Arnon Sol	D	7.8	8.3	8.5	8.6	8.6
	E	7.8	8.4	8.6	8.7	8.7
	F	7.8	8.3	8.5	8.6	8.6
	Control	7.8	8.0	8.4	8.5	8.6
T3 N and P	G	7.2	7.4	7.5	7.5	7.5
	H	7.2	7.3	7.4	7.4	7.4
	I	7.2	7.3	7.4	7.4	7.4
	Control	7.2	7.3	7.4	7.4	7.5

Table 3.
Height of E. crassipes reduced during the experimental period of 75 days.

are the primary ingredients for the natural growth of plants, so in the absence of these nutrients in the T3 group, slow growth and lower weight of water hyacinth were achieved.

The stress caused by constant herbivory allows energy to be diverted toward the development of daughter plants and new tissues, thereby reducing the overall growth rate of the plants and their sexual reproduction. Byrne et al. [30] stated that water hyacinth density and its spreading capacity were mainly related to asexual reproduction, so a decrease in reproductive capacities would reduce the expansion of mats and invasive potential of the water hyacinth.

In this study, it was observed that young hyacinths were controlled more rapidly than older plants that were confirmed by the studies of Ray et al. [31]. They reported that adult weevils are attracted to young plants because of the presence of some volatile substance that encourages them to feed especially at the previous site of injury but as the age increases it decreases, taking a longer time to be controlled by the weevils. So, managing larger plants is a tedious job in natural conditions that can only be made successful by releasing high inoculation loads of weevils, which increases production of smaller leaves and helps to destroy more leaves.

The change in fresh weights from 3.3% to 40.21 % in our study was very similar to the losses (−5 to −50%) reported by Del fosse and Cullen [32]. Tipping et al. [6] reported that weevil herbivory leads to a 50% reduction in biomass and inflorescence, but had less effect on the coverage area. Firehun [22] reported that three pairs of *N. bruchi* reduced 30% production of ramets, new leaves, and biomass thus reducing the productive capacity and vigor of the water hyacinth.

In the T2 group, due to high nutrient content (1.6 mg l^{-1} N and 1.0 mg l^{-1} P), the mass production of water hyacinths was achieved. In the control group, the wet weight obtained during the 10 weeks of the trial was almost twice as high as treatment (T2), but insect feeding still slightly affected (2.7–4.8%) the plant growth. Heard and Winterton [23] achieved greater damage at higher nutrient concentrations due to the greater production of offspring, that is a high reproductive rate (93 times in a generation). Due to the high reproductive rate, damage by weevils magnify over generations. However, the present findings contrasted with the study of Heard and Winterton, as the study was restricted to only one generation, hence less damage was achieved.

The collapse of the water hyacinth population by weevils in natural water bodies takes a time from 14 months to 24–36 months or 6 years [7, 24, 33]. The reason for this might be that water hyacinth growth and reproduction occur at a more rapid rate than the weevil's growth rate, so the weevils take a longer time to bring down the population of weeds.

5. Conclusion

Thus, the assessment of optimum densities of the weevil in different nutrients decreases the growth of its host plant up to 3.3–40.21% in 75 days except for extremely eutrophic conditions. The impact of these insects is supposed, to be evident as sudden, widespread eradication of the macrophyte, but rather it occurs gradually in our studies through slight changes in the phenology, morphometry, and productivity in the *Eichhornia's* population. We emphasize the need to treat the weed infestations as soon as the growth begins, usually before the plants start flowering, to minimize seed production. Thus, the outbreak of this aquatic weed could be sustainably managed by the judicious use of this potential Ecosystem Engineer. The biological control may act as flawless standalone technique for the

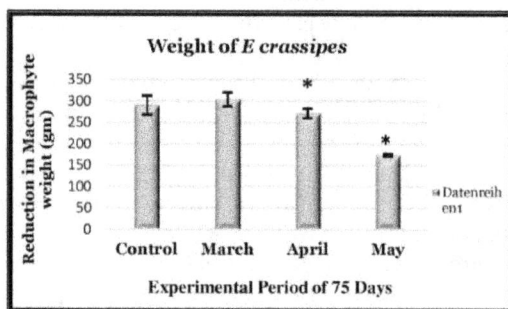

Figure 2.
Weight of E. crassipes was reduced during the Natural Condition of Experiment (T1) group after 75 days of inoculation (n = 10). The bars in the data represent the means, and the error bars are the standard error. The (*) over the month of April and May showed significant differences (p < 0.05) when compared to the control group, according to the Tukey HSD test.

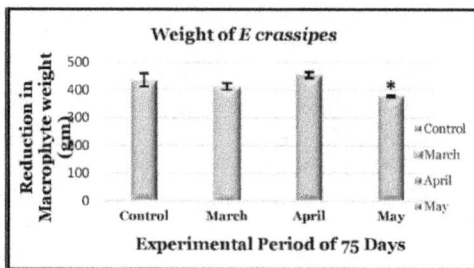

Figure 3.
Weight of E. crassipes was reduced in the presence of Hoagland and Arnon's solution (T2) group after 75 days of inoculation (n = 10). The bars in the data represent the means, and the error bars are the standard error. The (*) over the month of May showed non-significant differences (p > 0.05) when compared to the control group, according to the Tukey HSD test.

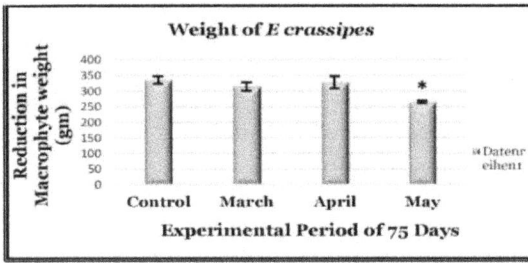

Figure 4.
Weight of E. crassipes was reduced in the absence of Nitrates and Phosphate content from Hoagland and Arnon's solution (T3) group after 75 days of inoculation (n = 10). The bars in the data represent the means, and the error bars are the standard error. The (*) over the month of May showed significant differences (p < 0.05) when compared to the control group, according to the Tukey HSD test.

control of water hyacinth. However, it should always be noticed that this technique is not always targeted at eradicating but, rather, it aims at managing populations to a level of permanent stress, thus bringing an effective control in the long run. In any case, this study clearly proves the words of T.D. Center that "any number of weevils is better than none". So, emancipating *Neochetina bruchi* from our natural water bodies to control this specific weed is a wise approach (**Figures 2–4**).

| *Neochetina bruchi* | *Scars after feeding of the biological control agent* | *Black scars after the insect herbivory* |

Author details

Prerna Gupta[1*] and Sadhna Tamot[2]

1 Laboratory of Endocrinology, Department of Biosciences, Barkatullah University, Bhopal, India

2 Department of Zoology, Sadhu Vaswani Autonomous College, Bhopal, India

*Address all correspondence to: prernag1707@gmail.com

IntechOpen

References

[1] Gecheva G, Yurukova L, Cheshmedjiev S. Patterns of aquatic macrophyte species composition and distribution in Bulgarian rivers. Turkish Journal of Botany. 2013;37:99-110

[2] Abobi SM, Yehoah AA, Kpodonu TA, Alhassan EH, Abarike ED, Atindaana SA, et al. Socio-ecological importance of aquatic macrophytes to some fishing communities in the Northern region of Ghana. Elixir Bio Diversity. 2015;79:30432-30437

[3] Kodarkar M. Bhoj Wetland Experience and Lesson Learned in Brief. Bhopal, India: Madhya Pradesh Lake Conservation Authority; 2006

[4] Bhat SP, Ramachandra TV. Macrophyte diversity in relation to water quality of Bangalore lakes. In: Conference Paper of Lake 2014 Conference on Conservation and Sustainable Management of Wetland Ecosystems in Western Ghat; Sirsi, Central Western Ghats, EWRG, Indian Institute of Science, India. 2014

[5] Gupta P, Tamot S, Shrivastava VK, Chakarde R. Seasonal variations in diversity of aquatic macrophytes of Upper lake Bhopal. Ecology Environment Conservation. 2020;26(8):231-235

[6] Tipping P, Martin M, Pokorny E, Nimmo K, Fitzgerald D, Dray AF, et al. Current levels of suppression of water hyacinth in Florida USA by classical biological control agents. Biological Control. 2014;71:65-69

[7] Goyer RA, Stark JD. The impact of *N, eichhorniae* on water hyacinth in Southern Lousiana. Journal of Aquatic Plant Management. 1984;22:57-61

[8] Ramchandra TV. Need for Conservation and Sustainable Management of Wetlands. Bangluru,

India: Energy & Wetlands Research Group, Centre for Ecological Sciences, Indian Institute of Science; 2010

[9] Ghosh D. Water hyacinth-Befriending the noxious weed [Feature Article]. Science Reporter. 2010. p. 48

[10] Naseema A, Praveena R, Nair R, Peethambaran C. *Fusarium pallidoroseum* for management of water hyacinth. 2004;86(6):770-771

[11] Gore P. Management of water hyacinth *E. crassipes*. Mart, Solms through bio-control agents with special reference to *Neochetina* spp, at Raipur district (thesis). Department of Entomology Indira Gandhi Krishi Vishwavidyalaya Raipur Chhattisgarh; 2017

[12] Julien MH, Griffiths MW, Wright AD. Biological control of water hyacinth—The weevils *N. bruchi* and *N. eichhorniae* Biologies host ranges and rearing releasing and monitoring techniques for biological control of *E. crassipes*. ACIAR Monograph. 1999:60-87

[13] Haller WT, Gettys LA, Bellaud M. Biology and Control of Aquatic Plants—A Best Management Practices. Gainesville, Florida, USA; 2009. pp. 41-46

[14] Kumar S. Biological based chemical integration for early control of water hyacinth. Indian Journal of Weed Science. 2011;43:211-214

[15] Cuda JP. Introduction to biological control of aquatic weeds. In: Haller WT, Gettys LA, Bellaud M, editors. Best Management Practices Manual for Aquatic Plants. Marietta GA: Aquatic Ecosystem Restoration Foundation; 2009. pp. 47-54

[16] Datta S. Aquatic Weeds and Their Management for Fisheries. CIFE Centre,

Salt Lake City Kolkata West Bengal India; 2009. pp. 1-22

[17] Netherland MD. Chemical control of aquatic weeds. In: Biology and Control of Aquatic Plants a Best Management Practices Handbook. Florida, USA: Aquatic Ecosystem Restoration Foundation Gainesville; 2009. pp. 65-77

[18] Raphael A. Biological control of invasive weed species nigerian experience. International Journal of Agriculture Research. 2010;**5**:121100-121106

[19] Lancar L, Krake K. Aquatic weeds and their Management. International Commission on Irrigation and Drainage. 2002

[20] Center TD. Biological control of weeds, water hyacinth and water lettuce. In: Rosen D, Bennett FD, Capinera JL, editors. Pest Management in the Subtropics. Biological Control—A Florida Perspective. UK: Intercept Ltd; 1994. pp. 481-521

[21] Center TD, Van TK, Dray FA, Franks SJ, Rebelo MT, Pratt PD, et al. Herbivory alters competitive interactions between two invasive aquatic plants. Biol Controlled. 2005;**2005**:33173-33185

[22] Firehun Y, Struik P, Lantinga EA, Taye T. Pre-release evaluation of *Neochetina* weevils potential for the management of *E. crassipes* [Mart.] Solm in the rift valley of Ethiopia. Academic Journal Agriculture Research. 2016;**47**:394-403

[23] Heard T, Winterton SL. Interactions between nutrient status and weevil herbivory in the biological control of water hyacinth. Journal of Applied Ecology. 2000;**37**:117-127

[24] Kumar S. Aquatic weeds problems and management in India. Indian

Journal of Weed Science. 2011:43118-43138

[25] Wilson JR, Hill G, Rees M, Holst N. Water hyacinth population dynamics. In: Proceedings of the Second Meeting of the Global Working Group for the Biological and Integrated Control of Water Hyacinth; Beijing, China. Australian Centre for International Agricultural Research (ACIAR); 2000. p. 152

[26] Gopal B. Water Hyacinth. Amsterdam: Elsevier; 1987

[27] Lambers H, Stuart F, Pons T. Plant Physiological Ecology. New York: Springer; 2008. p. 604

[28] Venter N, Hill MP, Hutchinson SL, Brad SR. Weevil borne microbes contribute as much to the reduction of photosynthesis in water hyacinth as does herbivory. Biological Control. 2013;**64**:138-142

[29] Charudattan R, Perkins BD, Littell RC. The effects of fungi and bacteria on the decline of arthropod damaged water hyacinth in Florida. Weed Science. 1978;**26**:101-107

[30] Byrne MMP, Hill M, Robertson A, King A, Jadhav N, Katembo J, et al. Integrated management of water hyacinth in South Africa development of an integrated management plan for water hyacinth control combining biological control herbicidal control and nutrient control tailored to the climatic regions of South Africa. Water Research Commission Report. 2010;**454**(10):302

[31] Ray P, Kumar S, Pandey AK. Impact evaluation of *Neochetina* on different growth stages of water hyacinth. Journal of Plant Protection Research. 2009;**49**(1):7-13

[32] Delfosse ES, Cullen JM. New activities in biological control of weeds in Australia. II. Echium plantagineum:

Curse or salvation? In: Proceedings of V International Symposium Biological Control of Weeeds Brisbane, Australia; 1980. pp. 563-574

[33] De Loach CJ, Cordo HA. Ecological studies of *Neochetina bruchi* and *N. eichhorniae* on water hyacinth in Argentina. Journal of Aquatic Plant Management. 1976;**14**:53-59

www.ingramcontent.com/pod-product-compliance
Lightning Source LLC
Chambersburg PA
CBHW081539190326
41458CB00015B/5598